NO_x Emission Controls for Heavy-Duty Vehicles: Toward Meeting a 1986 Standard

Final Report of the

Motor Vehicle Nitrogen Oxides Standard Committee
Assembly of Engineering
National Research Council

NATIONAL ACADEMY PRESS
Washington, D.C. 1981

University of Charleston Library
Charleston, WV 25304

NOTICE: The project that is the subject of this report was approved by the Governing Board of the National Research Council, whose members are drawn from the councils of the National Academy of Sciences, the National Academy of Engineering, and the Institute of Medicine. The members of the Committee responsible for the report were chosen for their special competences and with regard to appropriate balance.

This report has been reviewed by a group other than the authors according to procedures approved by a Report Review Committee consisting of members of the National Academy of Sciences, the National Academy of Engineering, and the Institute of Medicine.

The National Research Council was established by the National Academy of Sciences in 1916 to associate the broad community of science and technology with the Academy's purposes of furthering knowledge and of advising the federal government. The Council operates in accordance with general policies determined by the Academy under the authority of its congressional charter of 1863, which establishes the Academy as a private, nonprofit, self-governing membership corporation. The Council has become the principal operating agency of both the National Academy of Sciences and the National Academy of Engineering in the conduct of their services to the government, the public, and the scientific and engineering communities. It is administered jointly by both Academies and the Institute of Medicine. The National Academy of Engineering and the Institute of Medicine were established in 1964 and 1970, respectively, under the charter of the National Academy of Sciences.

The study culminating with this report was performed under contract number 68-01-6188 between the U.S. Environmental Protection Agency and the National Academy of Sciences.

Library of Congress Catalog Number   81-85575

International Standard Book Number   0-309-03226-1

Available from

NATIONAL ACADEMY PRESS
2101 Constitution Avenue, N.W.
Washington, D.C.   20418

Printed in the United States of America

PREFACE

In late 1980 the U.S. Environmental Protection Agency (EPA), as part of its rulemaking in establishing emission standards for heavy-duty vehicles called for in the Clean Air Act as amended 1977 (42 USC 7401 et seq.) requested that the National Research Council study the technological feasibility of meeting a more stringent standard for nitrogen oxides (NO_x) emissions by 1986. The Assembly of Engineering established the Vehicle Nitrogen Oxides Standard Committee to carry out the requested study.

The committee investigated NO_x emission control technologies and the impacts of those technologies on heavy-duty engines, cost, efficiency, performance, and emissions other than NO_x. In pursuing its charge, the committee solicited information from interested parties in the United States, Europe, and Japan. Committee members and staff visited the EPA Emissions Control Laboratory in Ann Arbor, Michigan, and the headquarters of the General Motors Corporation and the Ford Motor Company in Detroit. Several domestic and foreign engine manufacturers made presentations to the committee at its January 1981 meeting in Chicago. (Appendix A summarizes the committee's contacts with outside sources of information.)

In its search for data, the committee mailed a questionnaire (Appendix B) to domestic and foreign engine and parts manufacturers. In addition, contacts were made with foreign manufacturers and other companies at the February 1981 Meeting of the Society of Automotive Engineers. EPA, the U.S. Departments of Energy and Transportation, the California Air Resources Board, and government and private research laboratories provided additional data.

The policy background of the study underwent continual change. EPA had previously determined that a 1.7-gram per brake horsepower hour duty NO_x emissions would be required by the Clean Air Act. The agency was expected to publish a notice of proposed rulemaking on this issue in late 1980. Instead, it published an advance notice of proposed rulemaking, in the January 19, 1981, Federal Register, in which it stated the opinion that diesel engines could not attain the 1.7-g/bhp-h standard. The hearing dates specified in the advance notice were postponed from mid-summer to late fall.

The installation of a new administration in the White House and a change of leadership in EPA provided further changes in the public policy arena. In April 1981 the White House distributed a document outlining President Regan's proposed "Actions to Help the U.S. Auto Industry." On April 13, 1981, EPA published in the <u>Federal Register</u> its plan for implementing the President's program. In that plan EPA announced several changes affecting emission regulations for heavy-duty vehicles, including its intent to promulgate an NO_x standard based on the NO_x emission level that can be achieved by diesel engines. The target date for a notice of proposed implementation is June 1982.

While EPA and the Executive Branch are instituting changes in regulatory policy on emission control, Congress is holding hearings (as of mid-1981) on the Clean Air Act and its 1977 amendments. Emission standards for heavy-duty vehicles are one of the issues that will be discussed in these hearings.

Some of the regulatory changes proposed by EPA have to do with engine certification test procedures, which have a bearing on our study. We have noted these where it is appropriate. Our proposed changes in policy do not affect the committee's findings, because its main focus was on the available technology and the impacts of this technology's use.

The committee held its first meeting on December 5, 1980, and completed its draft report by the first week of May 1981. In order to carry out its full charge with the time and resources available, the committee devoted most of its time to its main charge of technology assessment. The committee noted the previous work of the NRC's forthcoming light-duty diesel study, which made a detailed examination of health and environmental impacts on the dollar value of the benefits of controls. Consequently, our committee did not study these topics in detail. We do, however, discuss them in the context of heavy-duty engine regulations.

Thus, the committee completed the tasks listed in the contract with EPA by focusing mainly on an appraisal of control technology, the area of expertise of most committee members. We sought to provide an independent analysis of the scientific, technical and cost issues raised by NO_x emission controls for heavy-duty vehicles of model year 1986. The committee noted several policy questions that arose from the technical findings in this study; these questions should be considered in a rulemaking procedure, but in the committee's judgment their answers require policy decisions beyond the scope of the committee's charge. One committee member disagreed with this judgment and has attached a statement of his views as Appendix E.

ACKNOWLEDGMENTS

This report was prepared by the Motor Vehicle Nitrogen Oxides Standard Committee under the chairmanship of Laurence S. Caretto. The committee itself is solely responsible for the report's findings and conclusions, but a number of others made important contributions to the committee's deliberations.

The committee was fortunate in having the able counsel of two advisors, William J. Lux, Manager, Product Engineering Center, John Deere Dubuque Works, and Russell P. Sherwin, Professor of Pathology, School of Medicine, Department of Pathology, University of Southern California.

The committee extends its thanks to those employees of the heavy-duty vehicle manufacturing industry who served as liaison to the committee or discussed the issues with committee members. They provided briefings, data, and materials that helped the committee understand the research and production issues confronting the industry.

Thanks are also given to the personnel of the Environmental Protection Agency who supplied the committee with materials and data. Their generous cooperation helped the committee to understand the difficulties and complexities of implementing the relevant sections of the Clean Air Act as amended 1977 (42 USC 7401 et seq.). The committee also received assistance from other federal agencies, most notably the Departments of Transportation and Energy.

The final preparation of the report was guided by comments from anonymous reviewers designated by the Assembly of Engineering, under the direction of its Executive Director, David C. Hazen. The committee is indebted to Duncan M. Brown, who served as editor to the committee.

The director of the study was Dennis F. Miller. Special recognition is due Vivian Scott, administrative assistant to the committee, and to Juliet W. Shiflet and Julia W. Torrence, secretaries, who served the committee with dedication.

MOTOR VEHICLE NITROGEN OXIDES STANDARD COMMITTEE

LAURENCE S. CARETTO, (Chairman), Professor, Department of Mechanical and Chemical Engineering, California State University, Northridge

ROY E. ALBERT, Professor, Institute of Environmental Medicine, New York University Medical Center

GEORGE R. HEATON, Jr., Research Associate and Program Manager, Center for Policy Alternatives, Massachusetts Institute of Technology

JOHN B. HEYWOOD, Professor, Department of Mechanical Engineering, Massachusetts Institute of Technology

CHARLES H. KRUGER, Jr., Professor, Department of Mechanical Engineering, Stanford University

EDWIN S. MILLS, Professor, Department of Economics, Princeton University

PHILLIP S. MYERS, Professor, Mechanical Engineering University of Wisconsin-Madison

HENRY K. NEWHALL, Manager, Fuels Division, Chevron Research Company

---

DENNIS F. MILLER, Study Director
VIVIAN SCOTT, Administrative Assistant
JULIET SHIFLET, Secretary
JULIA TORRENCE, Secretary

CONTENTS

EXECUTIVE SUMMARY	x

1 INTRODUCTION — 1

THE HEAVY-DUTY FLEET AND ITS USE PATTERNS	2
HEAVY-DUTY EMISSIONS REGULATIONS	3
Test Procedures	3
Proposed Emissions Standards	8
UNCONTROLLED ENGINE EMISSIONS AND CURRENT STANDARDS	8
CONTRIBUTION OF HEAVY-DUTY ENGINES TO TOTAL EMISSIONS	14
REGULATORY OPTIONS UNDER THE CLEAN AIR ACT	18
REFERENCES	20

2 HEAVY-DUTY DIESEL ENGINES

DESCRIPTION OF THE PROBLEM	21
Diesel Combustion and Emissions	21
Diesel Engine Trade-offs	23
CONTROL TECHNIQUES	24
Retarding Injection Timing	24
Varying the Shape of the Rate-of-Injection Curve	24
Turbocharging and Charge Cooling	26
Modifying Engine Designs	26
Exhaust Gas Recirculation (EGR)	26
Catalytic Controls	26
Water Injection	27
Turbocompounding	27
Insulating Engines Thermally	27
Trapping Particulates	27
Using Alternative Fuels	28
Use of Alternative Engines	28
1986 AVAILABILITY OF CONTROL TECHNIQUES	28
The System Problem	28
AVAILABLE DATA	29
ANALYSIS OF DIESEL NO_x EMISSION CONTROL DATA	33
SUMMARY	44
REFERENCES	46

3	HEAVY-DUTY GASOLINE ENGINES	47
	INTRODUCTION	47
	Control Techniques	47
	EFFECTS OF ENGINE MODIFICATIONS ON EMISSIONS AND FUEL ECONOMY	48
	Air/Fuel Ratio	48
	Spark Timing	48
	Exhaust Gas Recirculation	48
	Valve Timing	50
	Combustion Chamber Redesign	50
	Compression Ratio Changes	50
	Fuel System Modifications	50
	PERFORMANCE OF CURRENT HEAVY-DUTY GASOLINE ENGINES	50
	Examples of Current-Production Low-NO_x Engines	51
	EMISSIONS DATA FOR PROTOTYPE 1986 CONTROL SYSTEMS	57
	Noncatalytic NO_x Controls	57
	Catalytic Control of NO_x Emissions	60
	New Engine, Fresh Catalyst Emissions Data	63
	Catalyst Durability	66
	Overall Assessment of Catalytic NO_x Control	70
	CONCLUSIONS	70
	REFERENCES	72
4	ENVIRONMENTAL HEALTH EFFECTS	73
	BIOMEDICAL EFFECTS OF NO_x	73
	HEALTH ASPECTS OF POLLUTANT TRADE-OFFS	75
	SUMMARY AND CONCLUSIONS	76
	REFERENCES	77
5	CONTROL COSTS AND OTHER REGULATORY QUESTIONS	79
	CONTROL COSTS	79
	INDIRECT COSTS OF NO_x CONTROL	81
	REGULATORY ISSUES	85
	Differences Between Gasoline and Diesel Engines	85
	Vehicle Size Considerations	86
	Emissions Averaging	87
	Test Procedures	88
	Regulations and Technological Feasibility	89
	TECHNOLOGY BEYOND 1986	90
	Improved Fuel Controls	91
	Catalyst Technology	91
	Exhaust Gas Recirculation	91
	Particulate Traps	92
	Alternative Fuels and Engines	92
	CONCLUSIONS	94
	REFERENCES	95

APPENDIX A:	SUMMARY OF COMMITTEE CONTACTS	97
APPENDIX B:	LETTER AND QUESTIONNAIRE	99
APPENDIX C:	CERTIFICATION TEST CYCLES FOR HEAVY-DUTY ENGINES	104
APPENDIX D:	GLOSSARY OF TECHNICAL TERMS	110
APPENDIX E:	STATEMENT OF GEORGE R. HEATON, JR	112

EXECUTIVE SUMMARY

Heavy-duty vehicles are extremely varied in size, operating conditions, and vehicle type; the category includes trucks, buses, large vans, and recreational vehicles. Gasoline engines, with relatively high fuel costs and low capital costs, dominate the lighter end of the category, powering most vehicles with gross vehicle weight ratings between 8,500 and 26,000 pounds. Diesel engines dominate the heavier end, with gross vehicle weight ratings exceeding 26,000 pounds.

It has been estimated that heavy-duty vehicles contribute about 10 percent of the nation's total emissions of nitrogen oxides (NO_x). In the California South Coast Air Basin, their contribution to total NO_x emissions is estimated to be about 18 percent. Both the U.S. Environmental Protection Agency (EPA) and the California Air Resources Board predict that, because of reductions in NO_x emissions from other sources, the relative importance of NO_x emissions from heavy-duty vehicles will increase over the next decade.

The Clean Air Act as amended 1977 (42 USC 7401 et seq.) required EPA to promulgate an emission standard for NO_x, effective in the 1986 model year, that represents a 75-percent reduction in emissions from a baseline corresponding to the uncontrolled NO_x emissions of heavy-duty gasoline engines. Compliance will be determined by using the new heavy-duty "transient" test procedure, which replaces the old, "steady-state" procedure.

EPA has determined that the baseline NO_x emission level of uncontrolled heavy-duty gasoline engines, measured by the transient test procedure, is 6.8 grams per brake horsepower-hour (g/bhp-h). Thus the statutory 75-percent reduction in emissions would result in a standard of 1.7 g/bhp-h.

Heavy-duty diesel engines without emission controls have NO_x emission levels one and one-half to three times as high as the uncontrolled gasoline engine baseline. The percentage reduction required for diesel engines to meet the same standard as gasoline engines is therefore greater than 75 percent, at between 83 and 92 percent. Because of EPA's uncertainty about whether the 1.7-g/bhp-h standard can be met by production engines in 1986, the agency has recently suggested, for 1986, an interim standard of about 4 g/bhp-h.

EPA, in addition to proposing a new transient test cycle to replace existing steady-state test cycles, has proposed that the emission standards apply to engines until the ends of their useful lives (rather than until the end of a specified period of operation), and has introduced a new requirement that a specified fraction of production engines must meet the standard in an audit of production line engines. New engines, on the average, will need emission levels lower than that set by the numerical standard in order to comply with these requirements.

Major Findings

I. General Finding

Data from heavy-duty vehicle engines with NO_x emission levels significantly below those of today's engines are limited. In addition, only a small fraction of these data were gathered using the transient test procedure. Consequently, there is major uncertainty about the level of NO_x emission standard that can be achieved in the mid-1980s, and about the effects that a given standard will have on engine emissions of hydrocarbons, carbon monoxide, and particulates; fuel consumption; performance; and durability.

II. Findings on Diesel Engines

o Current 49-state production diesel engines have already achieved NO_x reductions of 20 to 60 percent from the uncontrolled diesel NO_x emission level; diesel engines sold in California have achieved 50- to 75-percent reductions in NO_x emissions relative to the uncontrolled diesel levels.

o The additional control techniques available to meet the 1986 NO_x standard for heavy-duty diesels are modifications in fuel injection timing and fuel injection pressure, cooling of the intake air charge, and possibly exhaust gas recirculation (EGR). Electronic systems to control timing are expected to be on the market by 1986, but whether they will have been available long enough to be incorporated widely in production engines is uncertain, as is the extent of their advantages in emissions control. Charge cooling using ambient air will be available, since it requires mainly straightforward hardware changes. The outlook for exhaust gas recirculation is not clear; it may be possible for some engine manufacturers to recirculate small amounts of exhaust gases in their engines without significant losses of durability and driveability, especially if electronic control systems are available. It appears unlikely that substantial amounts of exhaust gas can be recycled satisfactorily in all types of heavy-duty diesel engines.

- With currently available emission control systems and those under development, lowering NO_x emissions raises hydrocarbon and particulate emissions, increases fuel consumption, and decreases engine performance. (Engine-out particulate emissions might be controlled by particulate traps, but the availability of such devices in 1986 is uncertain.)

- The trade-offs between NO_x emissions and particulate and hydrocarbon emissions, and between NO_x emissions and fuel consumption for relatively new engines, have been estimated by the committee as follows:

Engine-Out Emissions (g/bhp-h)			Fuel Consumption
NO_x	Particulates	Hydrocarbons	Penalty, percent
8	0.4-0.5	0.6-0.8	0
6	0.5-0.7	0.7-1.4	2.5-4
4	0.6-1.0	0.8-1.7	7-12

These values are based on the new transient test cycle. For each level of emissions control, the fuel consumption penalty is defined as the increase in fuel consumption relative to the fuel consumption that an engine with the same technology would have at an NO_x emission level of 8 g/bhp-h.

- The NO_x standards corresponding to the above low-mileage emission levels would be higher than the low-mileage levels, after allowance is made for deterioration in emission control over the engine's useful life, and for production line variability.

- There are no data that show the technical feasibility of the heavy-duty diesel engine's meeting an NO_x emission standard of 1.7 g/bhp-h.

III. Findings on Gasoline Engines

- Control techniques available to meet the 1986 heavy-duty engine NO_x standard are based on engine modifications and the use of catalytic converters for exhaust treatment. Engine modifications include changes in air-fuel ratio, spark timing, and valve timing; redesign of combustion chambers, intake manifolds, and fuel metering systems; and use of EGR. Three-way catalyst systems, often with feedback-controlled fuel metering, can provide additional NO_x reductions in the exhaust system.

o   Without the use of catalysts in the exhaust system to reduce emissions of NO_x,

   -- NO_x emission levels of about 5 g/bhp-h can be achieved with little or no effect on fuel consumption.

   -- NO_x levels of about 3 g/bhp-h can be achieved with a 3- to 7-percent increase in fuel consumption.

   -- Attempts to engineer emission controls to achieve NO_x levels lower than about 3 g/bhp-h without catalysts will increase fuel consumption and degrade engine performance substantially.

o   With three-way catalytic systems using fresh catalysts, NO_x emission levels below the 1.7-g/bhp-h standard mandated by the Clean Air Act as amended 1977 (42 USC 7401 et seq.) have been achieved by relatively new engines, with 4- to 7-percent increases in fuel consumption.

o   No data on the durability of catalytic NO_x control systems on heavy-duty gasoline engines are available. However, data on heavy-duty engine exhaust temperatures suggest the potential for significant deterioration of catalysts. Catalytic NO_x control systems cannot be considered practical until adequate durability data become available.

IV. Other Findings

o   From a public health standpoint, it might be imprudent to suppress the emissions of NO_x from heavy-duty vehicles at the cost of a substantial increase in the emission of particulates. The extent to which these two species affect human health requires further study, especially to assess the carcinogenic risk of diesel particulates.

o   For an increase in fuel consumption greater than about 2 percent, the most significant cost of further NO_x control will be the cost of the increased fuel consumption. For larger fuel penalties, the increased fuel cost will be substantially greater than the initial cost of the control system.

o   This report has addressed technology available for use in 1986. Beyond that time technology that is now uncertain can be further developed, and new technologies may emerge. These technological improvements will be influenced by regulatory requirements.

Concluding Comment

The above findings are based on the committee's review of the available technical information. The committee has not performed a complete regulatory analysis and has not commented on the policy implications of its results. The committee did note that its findings suggest the possibility of different emissions standards for different engine types (diesel and gasoline) or different vehicle sizes. The committee did not analyze all of the policy consequences, but simply raised the issue as one that should be addressed in considering NO_x standards for heavy-duty vehicles.

The findings of the committee outline, with the appropriate ranges of uncertainty, some consequences of various levels of NO_x control. The major part of this study addressed the costs of NO_x control in terms of both dollars and increased emissions of other species; these costs increase as the standards become more stringent. We have not been able to quantify the benefits of NO_x control in heavy-duty vehicles, but these should also increase with more stringent controls, provided increases in emissions of other species can be minimized.

It is the task of the Congress and the Environmental Protection Agency to determine whether the benefits of a given level of NO_x control justify the costs involved. We hope that this report is helpful.

Chapter 1
INTRODUCTION

Emissions regulations for heavy-duty vehicles set limits on releases of hydrocarbons, carbon monoxide, oxides of nitrogen (NO_x), and visible smoke. (The U.S. Environmental Protection Agency (EPA) has proposed changes in heavy-duty regulations that would replace visible smoke limits by limits on particulate emissions.)

A heavy-duty vehicle is defined by EPA as one with a gross vehicle weight rating (GVWR) of 8,500 pounds or above. Both diesel-powered and gasoline-powered engines are used as heavy-duty engines, though the heaviest vehicles are almost all equipped with diesel engines. The heavy-duty vehicle market consists of a wide variety of vehicles, including trucks, buses, large vans, and recreational vehicles. The manufacturers of many of these vehicles do not manufacture their own engines, but instead purchase engines from others. A given engine may thus have a variety of applications in many different vehicles.

To account for the variety of the heavy-duty vehicle fleet, the applicable emissions regulations are based on the output of the engine itself rather than on the performance of the vehicle. That is, standards are set in terms of the permissible amounts of emittants per unit of energy output (i.e., grams per brake horsepower-hour). This is in contrast to the light-duty vehicle regulations, which limit the various emissions to specified amounts per unit distance of vehicle travel (i.e., grams per mile).

The Clean Air Act as amended 1977 (42 USC 7401 _et seq._) set emission reduction targets for gaseous emissions from heavy-duty engines. For NO_x the target is a 75-percent reduction from a baseline representing uncontrolled gasoline engine emissions. For hydrocarbons and carbon monoxide, the limit is a 90-percent reduction from the uncontrolled gasoline engine baselines. The specification of the uncontrolled gasoline engine's emissions as the baseline for all engines makes the requirements for hydrocarbons and carbon monoxide stricter for gasoline engines than for diesels, because uncontrolled diesels have lower emissions of these species than gasoline engines. On the other hand, the NO_x reduction is a more difficult requirement for diesels, because uncontrolled diesel engines emit more NO_x than uncontrolled gasoline engines.

EPA has published an advance notice of proposed rulemaking (U.S. Environmental Protection Agency, 1981a), interpreting the 75-percent emissions reduction target for NO_x as requiring an NO_x emission level of 1.7 g/bph-h, as measured on the new heavy-duty "transient" test procedure (described later in this chapter and more fully in Appendix C). Associated with this emissions level are new definitions of useful life and new enforcement audit requirements, described in a later section of this chapter, "Regulatory Options Under the Clean Air Act."

This committee has investigated the available technology for reducing the NO_x emissions of heavy-duty vehicles, has evaluated the technological feasibility of meeting various possible NO_x standards, and has assessed the corresponding impacts on fuel economy, costs, and emissions of other species. In carrying out this study, the committee has reviewed recent technical literature and conducted a number of interviews and site visits, as outlined in Appendix A. The time available has limited the amount of information the committee was able to review in detail. Also, in many important areas data are simply unavailable. Many of our conclusions are therefore tentative, indicating the need for additional studies. Other conclusions have been expressed with large ranges of uncertainty.

THE HEAVY-DUTY FLEET AND ITS USE PATTERNS

The heavy-duty vehicle industry typically uses gross vehicle weight ratings (GVWR) as a basis for reporting production and sales data. The traditional industry categories are as shown in Table 1. EPA's heavy-

TABLE 1 Vehicle Weight Classes Used By The U.S. Vehicle Industry

Class	GVWR (lb)
1	0-6,000
2	6,001-10,000
3	10,001-14,000
4	14,001-16,000
5	16,001-19,500
6	19,501-26,000
7	26,001-33,000
8	33,001 and over

duty vehicle classification, including all vehicles with GVWRs over 8,500 pounds, thus encompasses some vehicles in industry's class 2 and all of classes 3 through 8.

Emission control for heavy-duty vehicles is complicated by the heterogeneity of the heavy-duty vehicle population, the diversity of engines and engine-transmission combinations, and the wide range of operating conditions and patterns of use. The buyer typically specifies the type of vehicle and the type and manufacturer of the engine, transmission, and other vehicle equipment. The choice of a gasoline or diesel engine is determined primarily by economics. Diesels have better fuel economy in terms of miles per gallon but higher initial costs. If the higher cost of the diesel engine can be recovered from fuel savings in something like three years, the buyer will choose a diesel unless there are compelling reasons not to do so.

Table 2, taken from the EPA draft regulatory analysis, (U.S. Environmental Protection Agency, 1980a), shows heavy-duty vehicle sales by weight class and by year. The growth in the 8,501-10,000 pound class, and the decrease in classes 3, 4, and 5 (and perhaps 6) is evident. Table 3, adapted from the same source, shows the economic judgment already being made in favor of the diesel engine in classes 7 and 8, and probably in class 6. Most current projections of diesel use in heavy-duty vehicles (e.g., Jambekar and Johnson, 1981) forecast increasing use of diesel engines in the lighter heavy-duty weight classes. The emissions projections in this chapter and the cost projections in Chapter 5 use these projections. This shift to diesels would result in an increase in emissions without additional controls.

Heavy-duty trucks operate in both rural and urban areas. The proportion of use in each is important, because rural areas have better air quality than urban areas, and correspondingly less need to control emission. Table 4 shows that the heaviest trucks (in classes 7 and 8), almost all of which are diesel-powered (Table 3), drive most of their miles in rural areas. With the information in Tables 3 and 4, one can calculate roughly the relative amounts of pollutants emitted in rural and urban areas by heavy-duty trucks if one knows the amounts of pollutants emitted by the various classes of such vehicles. This latter can be deduced roughly from the amounts of fuel used by the different classes; in 1973, class 3 trucks consumed 2.6 percent of total highway fuels; class 4, 0.19 percent; class 5, 1.5 percent; class 6, 3.1 percent; class 7, 1.2 percent; and class 8, 9.9 percent. One can conclude from all of this that about half of the pollution produced by gasoline-powered heavy-duty vehicles was emitted in urban areas, while only about one-sixth of the heavy-duty diesel emissions were produced in urban areas.

HEAVY-DUTY EMISSIONS REGULATIONS

Test Procedures

In the current steady-state test procedures for certifying the compliance of heavy-duty engines with emissions standards, the emissions are measured

TABLE 2 U.S. Domestic Factory Sales of Trucks and Buses Plus Imports from Canada, by Gross Vehicle Weight Rating

Heavy-Duty Vehicles, by Class

Year	0–8,500[a]	8,501–10,000[a]	Class 3 (10,001–14,000)	Class 4 (14,001–16,000)	Class 5 (16,001–19,500)	Class 6 (19,501–26,000)	Class 7 (26,001–33,000)	Class 8 (33,000 and over)	Yearly Total	Vehicles Subject to Heavy-Duty Regulations
1979	2,557,153	148,829	17,366	2,361	3,147	139,542	47,239	173,675	3,089,312	532,159
1978	3,218,772	187,336	34,014	5,959	3,982	157,168	41,516	163,836	3,812,583	593,811
1977	2,972,752	173,017	30,064	3,231	4,989	160,396	32,249	148,728	3,525,426	552,674
1976	2,525,755	147,002	43,411	67	8,920	149,293	22,918	103,098	3,000,466	474,709
1975	1,790,355	104,201	19,497	6,508	13,916	152,070	24,698	74,896	2,186,141	395,786
1974	2,088,200	121,535	8,916	8,120	24,366	215,221	32,364	160,465	2,659,187	570,987
1973	2,370,208	137,949	52,558	8,744	37,043	199,481	40,816	155,814	3,002,613	632,405
1972	1,929,883	112,321	57,803	10,353	37,492	177,723	40,150	130,328	2,496,054	566,170

[a]The MVMA does not split sales at 8,500 pounds GVWR, but rather publishes sales for the 0–6,000 and the 6,001–10,000 pound classes. The split in the table represents EPA's estimate.

SOURCE: Adapted from U.S. Environmental Protection Agency, 1980 (from Motor Vehicle Manufacturers Association data).

TABLE 3 Diesel Factory Sales as a Percentage of All Heavy-Duty Vehicle Factory Sales

| Year | Sales by Class, as Percentages of Total Sales ||||||| All Heavy-Duty Vehicles |
	Class 2 (8,501-10,000)	Class 3 (10,000-14,000)	Class 4 (14,001-16,000)	Class 5 (16,001-19,500)	Class 6 (19,501-26,000)	Class 7 (26,001-33,000)	Class 8 (33,001 and over)	
1979	--	--	--	--	11	60	96	39
1978	--	--	--	--	8	62	96	32
1977	--	--	--	--	7	58	96	31
1976	--	--	--	--	4	49	94	24
1975	--	--	--	1	3	45	88	21
1974	--	--	--	--	2	40	88	28
1973	--	--	3	--	2	45	89	26
1972	--	--	2	--	2	32	89	24

SOURCE: Adapted from U.S. Environmental Protection Agency, 1980a.

TABLE 4 Urban Travel as a Fraction of Total Mileage, for Gasoline- and Diesel-Powered Vehicles

GVWR Class	Gasoline	Diesel
1	0.45	0.45
2A[a]	0.45	0.45
2B[b]	0.45	0.45
3	0.66	0.46
4	0.66	0.46
5	0.66	0.46
6	0.75	0.40
7	0.61	0.17
8	0.61	0.17

[a]Gross vehicle weight ratings 6,001–8,500 lb.

[b]Gross vehicle weight ratings 8,501–10,000 lb.

SOURCE: Adapted from Jambekar and Johnson, 1981.

while the engine is running on a dynamometer and operating at a defined series of steady loads and speeds. Engines must meet the standard after operation for 1,000 or 1,500 hours (for gasoline and diesel engines, respectively). EPA enforces emission standards for light-duty vehicles by production-line tests called selective enforcement audits (SEAs). In these audits, 40 percent of the engines tested are allowed to have emissions greater than the standard. This is called the acceptable quality level (AQL) requirement. Standards for heavy-duty engines are not currently enforced by such audits.

The certification test procedures for 1984 and beyond call for emission measurements while the engine is run through a "transient cycle," with second-by-second changes in speed and load (U.S. Environmental Protection Agency, 1980b). This cycle requires the installation of new dynamometers, capable of driving the engine to simulate conditions, such as downhill driving, which are not encountered in the steady-state tests. (Appendix C describes and compares the steady-state and transient test cycles.)

In addition, the new test procedures require engines to meet the standards until the ends of their "useful lives" (the engine manufacturer's suggested mileage before rebuilding for diesels.) Emission regulations for heavy-duty engines of model years 1984 and later contain a new 10-percent AQL requirement, so that 90 percent of the engines tested in a production line audit must meet the applicable standard. (EPA recently announced (U.S. Environmental Protection Agency, 1981b) plans to defer enforcement audits two years and to relax the AQL requirement to 40 percent.)

The AQL requirement and the point in the engine's lifetime when the standards apply are important factors in assessing control technology. Much of the data received by the committee are from new engines. Such data cannot be directly interpreted in terms of legal emissions standards. Emissions of new production engines must be lower than the standards with which they are intended to comply by a deterioration factor that accounts for any increase in emissions as the engines age and a safety factor to account for production line variability during enforcement audits. The deterioration factor is defined as the ratio of the emissions of the engine at the end of its useful life to those of the new engine. This factor is determined as part of the engine certification procedure for new engines. Once it is determined, subsequent emissions tests are made on new engines and the results are multiplied by the deterioration factor to yield a value for useful-life emissions. This value is the one that must meet the standard.

Another safety factor is used by manufacturers to ensure passing production line enforcement audits, based on the regulatory specification of the AQL and the manufacturer's estimate of production line variability. This factor decreases if production line variability decreases or if regulations are changed to allow the AQL (the percentage of engines that can fail the standard in a production line audit) to increase.

The effects of these various requirements can be summarized by a numerical example. Manufacturers are now required to meet a standard of

10 g/bhp-h for the sum of hydrocarbons and NO_x. With a deterioration factor of 1.25, a new engine must meet a standard of 10 divided by 1.25, or 8 g/bhp-h. If the manufacturer decides to set a 10-percent margin of safety to satisfy the proposed production-line audit requirement, engines would have to be designed to meet an emissions standard of 8 (0.1 x 8), or 7.2 g/bhp-h. Thus an evaluation of the technological feasibility of meeting this 10-g/bhp-h standard must look for a new engine capable of achieving an emission level of 7.2 g/bhp-h. Table 5 summarizes manufacturer and EPA estimates of new-engine emissions levels required to meet various NO_x standards.

Because of the relatively recent institution of the transient test cycle, the available data on engine emissions are derived largely from the old steady-state procedures. There is no general correlation between emissions measured on the two cycles; correlations are different for different emission species, for different engine types, and even for different manufacturers. The best correlations between the two cycles are those for NO_x emissions, but these correlations permit only estimates of transient-cycle emissions from measurements on the steady-state cycle.

The effect of the new test cycle on manufacturers' abilities to develop systems for meeting new emission standards within a given time is also important. For example, installing new dynamometer installations will require some time. It will also take time for manufacturers to develop data bases on the performances of their current and prototype engines, measured on the new transient cycle, and to evaluate the impact of a new definition of useful engine life.

Proposed Emissions Standards

The Clean Air Act as amended 1977 (42 USC 7401 et seq.) prescribed specific reduction targets for gaseous emissions and the "greatest degree of emissions reduction achievable" for particulate emissions. In carrying out these mandates, EPA has promulgated emissions standards for hydrocarbons and carbon monoxide to become effective in the 1984 model year. In addition, the agency has proposed standards for NO_x and particulate matter that would be effective in the 1986 model year. These are (in grams per brake horsepower-hour) 1.3 for hydrocarbons, 15.5 for carbon monoxide, 1.7 for NO_x, and 0.25 for particulates. These emissions are to be measured on the new heavy-duty transient test procedure and must be met by engines until the ends of their useful lives. At the time of this report, EPA is revising all these standards, and changing the 10-percent AQL requirement back to 40 percent (U.S. Environmental Protection Agency, 1981b).

UNCONTROLLED ENGINE EMISSIONS AND CURRENT STANDARDS

Current emission standards for heavy-duty engines regulate hydrocarbons, carbon monoxide, and oxides of nitrogen (NO_x). There are also opacity standards for

TABLE 5 Design Targets to Meet NO_x Standards Under EPA Proposed Test Procedures[a]

NO_x Standard (g/bhp-h)	Engine Type	Design Targets (g/bhp-h) as Estimated by:	
		EPA[b]	Manufacturers[c]
1.7	Diesel	1.19	1.2-1.4
1.7	Gasoline	1.01	1
4.0	Diesel	--	2.9-3.3
4.0	Gasoline	--	1.5-2.5[d]

[a]Transient test cycle; 10-percent AQL; standards to apply at end of engine's useful life.

[b]U.S. Environmental Protection Agency, 1980a.

[c]Indicated as a range or upper limit to preserve confidentiality.

[d]Estimates from data on other standards.

diesel smoke. As required by the Clean Air Act, EPA is proposing standards for diesel particulates, to become effective in the 1986 model year.

Table 6 gives estimates of emissions from uncontrolled heavy-duty engines. The significant points of this table are the differences between gasoline and diesel engines. Without emission controls, gasoline engines have much higher emissions of hydrocarbons and carbon monoxide than diesel engines, while diesel engines have higher emissions of NO_x. The committee estimates the uncontrolled NO_x emissions of diesel engines as one and one-half to three times as great as those of gasoline engines without NO_x controls.

Particulate emissions from gasoline engines are usually associated with lead salts from the lead additives used to improve octane numbers. Recent regulations limiting the lead content of fuel, have reduced emissions of lead particulates from gasoline-fueled vehicles. Diesel particulates are organic in nature, resulting from the diesel combustion process. Gasoline engines are generally considered insignificant sources of organic particulates. EPA has assumed that they can comply easily with any particulate standard and is not proposing that gasoline engine manufacturers run certification tests on particulate emissions.

Current and proposed emission regulations for heavy-duty engines are tabulated in Tables 7 (federal standards) and 8 (California standards). The levels shown for 1986 are only proposed levels, not actual regulations. The 1984 hydrocarbon and carbon monoxide standards are currently undergoing revision (U.S. Environmental Protection Agency, 1981b). In some cases standards for heavy-duty engines have been expressed as a limit on the sum of hydrocarbons plus oxides of nitrogen (HC + NO_x). This combined standard allows the manufacturers flexibility in designing engines to achieve reductions in these two species. More recently the standards have included limitations on both hydrocarbons and HC + NO_x. The ability to use this trade-off is limited by the hydrocarbon standard. If an engine with no hydrocarbon emissions were produced, it could have NO_x emissions of 10 g/bhp-h, according to the 1981 federal standards. If a 1981 federal engine just met the 1.5-g/bhp-h hydrocarbons standard, it would have an NO_x emissions limit of 8.5 g/bhp-h.

In evaluating the status of emissions control in heavy-duty engines, it is apparent that for gasoline engines, the main control problems are hydrocarbons and carbon monoxide, while for diesels, NO_x emissions will be most difficult to control. Quantitative comparisons are difficult because emissions as measured on the transient and steady-state cycles are not directly comparable and because data on uncontrolled diesel engines cover a wide range. For purposes of rough estimation, however, one may take the NO_x emissions of current federal diesel engines as 8 g/bhp-h on the transient cycle. The reduction from the initial, uncontrolled levels of 10-20 g/bhp-h shown in Table 6 thus corresponds to a 20- to 60-percent reduction. The California engines in current production have NO_x emissions of about 6 g/bhp-h on the transient cycle, corresponding to reductions of 40-70 percent.

TABLE 6 Emissions from Uncontrolled Heavy-Duty Engines

Engine Type	Test Cycle	Emissions Levels (g/bhp-h)				Reference
		Hydrocarbons	Carbon Monoxide	NO_x	Hydrocarbons + NO_x	
Gasoline	9-Mode	10.9	132	5.1	16	(a)
Diesel	13-Mode	1.3	5.6	10.5	11.8	(a)
Gasoline	Transient	13	155	6.8		(b)
Diesel[c]	Transient	1-3	3-6	10-20		(d)

[a]California Air Resources Board, 1976.

[b]Inferred from U.S. Environmental Protection Agency, 1980b, 1981a.

[c]Particulate emissions from 1978-1980 engines are 0.36 to 0.79 g/bhp-h on the transient cycle.

[d]Ranges based on information given to the committee by EPA and manufacturers.

TABLE 7 Federal Exhaust Emission Standards for Heavy-Duty Vehicles

Years	Gaseous Emission Limits g/bhp-h unless otherwise noted[a]				Smoke(% opacity)[b]
	Hydrocarbons	Carbon Monoxide	NO_x	Hydrocarbons plus NO_x	
Pre-1969	--	--	--	--	
1970-1973	275ppm	1.5%	--	--	40/20/--
1974-1978	--	40	--	16	20/15/50
1979-1983[c][d]	1.5	25	--	10	20/15/50
1984-1985[e][f]	1.3	15.5	10.7	--	20/15/50
1986 & Later[f]	1.3	15.5	[g]	--	[h]

[a] Prior to 1984 the steady-state procedure is used; for 1984 and later the transient test is specified. Units are grams per brake horsepower-hour unless otherwise noted.

[b] Smoke limits are on special cycle for measuring diesel smoke. The numbers are percent opacity limits for three conditions: acceleration, lug, and peak.

[c] Optional standard of 25 g/bhp-h carbon monoxide and 5 g/bhp-h hydrocarbons plus NO_x available for 1977-1983 in California and 1979-1983 federally.

[d] Changes in details of the test procedure made in 1979. Change in hydrocarbon analyzer gives higher measured hydrocarbons value by 0.5 g/bhp-h for same engine.

[e] Optional steady-state standards for 1984 diesels only are 0.5, 15.5 and 9 g/bhp-h for hydrocarbons, carbon monoxide, and NO_x, respectively.

[f] The U.S. Environmental Protection Agency (1981b) has announced that they will revise the 1984 hydrocarbons and carbon monoxide standards to a level that will not require catalysts on gasoline heavy-duty engines.

[g] The statutory Clean Air Act standard is 1.7 g/bhp-h; the U.S. Environmental Protection Agency (1981a) has announced its intention to propose a standard of approximately 4 g/bhp-h.

[h] The U.S. Environmental Protection Agency (1981c) has proposed a particulate standard of 0.25 g/bhp-h for 1986.

TABLE 8 California Exhaust Emission Standards for Heavy-Duty Vehicles

Years	Gaseous Emission Limits (g/bhp-h unless otherwise noted)			
	Hydrocarbons	Carbon Monoxide	NO_x	Hydrocarbons plus NO_x
Pre-1968	--	--	--	--
1969-1971	275 ppm	1.5%	--	--
1972	180 ppm	1.0	--	--
1973-1974	--	40	--	16
1975-1976	--	30	--	10
1977-1978[b]	1.0	25	7.5	--
1979[c]	1.5	25	7.5	--
1980-1983[c]	1.0	25	--	6
1984 and later[a]	0.5	25	--	4.5

[a] California has used the same test cycle as EPA. All numbers in this table are from the steady-state cycles. For 1984, manufacturers can choose the option of complying with the federal standards, using the transient cycle, except that the applicable NO_x standard is 5.1 g/bhp-h rather than 10.7 g/bhp-h. Units are g/bhp-h unless otherwise noted.

[b] Optional standard of 25 g/bhp-h for carbon monoxide and 5 g/bhp-h for the sum of hydrocarbons and NO_x are available for 1977-1983 in California and 1979-1983 federally.

[c] Changes in details of the test procedures were made in 1979. The change in the hydrocarbon analyzer raises the measured value of hydrocarbons by 0.5 g/bhp-h in a given engine.

The 1977 Clean Air Act amendments' requirement that all engines meet a common NO_x emissions standard, determined by taking a 75-percent reduction from the emissions levels of the average uncontrolled gasoline engine, means that the diesel engine must reduce NO_x emissions more than the gasoline engine. For the range of uncontrolled diesel emissions (one and one-half to three times the emissions of the average uncontrolled gasoline engine) the percentage reduction requirements for the diesel engine range from 83 to 92 percent.

CONTRIBUTION OF HEAVY-DUTY ENGINES TO TOTAL EMISSIONS

In 1977 the total national emissions of oxides of nitrogen ($NO + NO_2$) from man-made sources were about 2.2 million tons. Of this, about 10 percent came from heavy-duty vehicles. EPA estimates that, without further regulation of heavy-duty vehicles, this percentage will grow to 15 percent by 1990. The relative contribution of heavy-duty engines to total NO_x emissions depends on the location. In rural areas, without other significant sources of NO_x emissions, the relative contribution of the heavy-duty engine will be larger, but these will be areas of relatively low ambient pollutant concentrations.

Estimates of the total emissions of NO_x and of the relative contributions of heavy-duty engines of each type are displayed in Tables 9-11. Table 9 shows the nationwide inventory in 1977 as prepared by EPA using a modified emissions factor forecast, and an alternative inventory prepared by the Ford Motor Company. The inventories are in good agreement on total emissions and the contributions of different sources, though the EPA inventory assigns a greater contribution to heavy-duty engines than the Ford inventory. Table 10 tabulates EPA forecasts of an increase in both total emissions and heavy-duty emissions, but with the emissions from heavy-duty vehicles, assuming no further NO_x controls, increasing at a faster pace.

The emissions forecasts in Tables 9 and 10 are based on estimates of growth in population and industrial activity, the types of fuels to be used, and the emission control technology to be applied.

Table 11 shows the current emissions and forecasts for the Los Angeles (South Coast Air Basin) and San Francisco (Bay Area) air quality regions. Here the percentage contributions from heavy-duty vehicles are greater than the nationwide average. This is because controls on emissions from light-duty vehicles and stationary sources are more stringent in these regions than in the nation as a whole. Without further NO_x emissions standards for heavy-duty vehicles, heavy-duty vehicles are predicted to contribute 24 percent of the total NO_x inventory in the South Coast Air Basin by 1987. This is 44 percent of the total mobile source emissions inventory forecast for that year.

The ultimate impacts of emissions depend not only on the kinds of emissions that leave a mobile source, and their amounts, but also on their transport and the extent to which they are transformed chemically in the

TABLE 9 Estimates of the 1977 Inventory of NO_x Emissions from Heavy-Duty Vehicles

Source of Emissions	Estimates of Inventory			
	EPA[a]		Ford Motor Co.[b]	
	(kilotons per year)	(percentage of total)	(kilotons per year)	(percentage of total)
Heavy-duty vehicles:				
Gasoline	828	4	639	3
Diesel	1,506	7	1,412	7
Total mobile sources	9,454	43	9,124	42
Total all sources	21,775	100	21,736	100

[a]U.S. Environmental Protection Agency, 1980a.

[b]Information gathered on site visit, March 4, 1981. (See Appendix A.)

TABLE 10 U.S. Environmental Protection Agency Forecasts of Total National NO_x Emissions and the Percentage Contribution of Heavy-Duty Vehicles

Year	Total Emissions (Kilotons per year)	Contribution of Heavy-Duty Vehicles (percent)
1990	23,672-25,343	15
1999	28,037-31,390	16-17

SOURCE: U. S. Environmental Protection Agency, 1980a.

TABLE 11 California Air Resources Board NO_x Emissions Inventory (Current and Forecast) for South Coast Air Basin and Bay Area Air Quality Management District

Source of Emissions	Estimated and Forecast Emissions by Year			
	1979		1987[a]	
	(tons per day)	(percentage of total)	(tons per day)	(percentage of total)
South Coast Air Basin				
Heavy-duty vehicles:	245	18	274	24
Gasoline	(61)		(56)	
Diesel	(184)		(218)	
Total motor vehicles	830	60	622	54
Total all sources	1,380	100	1,158	100
Bay Area Air Quality Management District				
Heavy-duty vehicles:	126	16	139	19
Gasoline	(32)		(30)	
Diesel	(94)		(109)	
Total motor vehicles	417	55	317	44
Total all sources	765	100	720	100

[a] Assumptions: New heavy-duty engines controlled to hydrocarbon emission levels of 0.5 g/bhp-h and hydrocarbon-plus-NO_x emission levels of 4.5 g/bhp-h (measured on the steady-state cycles) in 1984 and later years. Light-duty engines controlled to NO_x emission levels of 0.4 g/bhp-h in 1983 and later years.

SOURCE: Informal communication, R. Bradley, California Air Resources Board, January 1981.

atmosphere. The impact of mobile source emission controls thus depends not only on the control technology used, but also on how the vehicles are operated. A 75-percent reduction in the NO_x emissions of heavy-duty vehicles cannot be translated directly into a change in atmospheric concentrations of any chemical species that results from NO_x emissions. Such a translation requires a knowledge of the emission source patterns for the region in question, the total atmospheric loading of emissions from all sources in the region, and the detailed transformation processes. Such knowledge is not available now. Although complete knowledge of all these processes is not available, it is possible to make good estimates of the effects of emissions control on air quality. For example, in regions where heavy-duty engines make a significant contribution to overall NO_x emissions, a significant reduction in the atmospheric burden of species formed from these emissions would be associated with reductions of heavy-duty NO_x emissions.

In evaluating the significance of these numbers it is important to consider the reduction in urban NO_x emissions required to attain and maintain air quality standards. It is further necessary to consider the cost of reducing a given amount of NO_x emissions from heavy-duty vehicles compared to the cost of the same reduction of NO_x emissions from other sources. The relationships between NO_x emission levels and those of other emissions should also be considered. This type of evaluation is beyond the scope of our report, but it should be an important ingredient in an overall regulatory program.

REGULATORY OPTIONS UNDER THE CLEAN AIR ACT

The Clean Air Act as amended 1977 (42 USC 7401 et seq.) is routinely referred to as "technology forcing" legislation, under which standards should be set in anticipation of technology's becoming available. The meaning of this term is less than clear. Nevertheless, the concept is legally sanctioned, and in specific instances the courts have upheld agency regulatory action based on the belief that technology "on the horizon" would be further developed to achieve compliance (Federal Court of Appeals, 1975).

The mobile source provisions in the original Act of 1970 differed significantly from most other pieces of regulatory legislation in assigning much less discretion to the regulatory agency, Congress chose, essentially, to write the emissions standards itself, specifying the required reductions in emissions in the legislation.

Although subsequent events have modified the 1970 standards and legislation somewhat, their basic purpose and structure have not been abandoned. Still, new concerns--energy conservation, international trade, cost and technological feasibility--pervade the regulatory process. These issues were brought to the fore in the 1977 amendments, which added the sections of the Act under which NO_x emissions of heavy-duty vehicles are to be controlled. In an important sense, the structure of the 1970 Act is preserved; Congress has itself specified a goal (a 75-percent reduction of NO_x emissions from their uncontrolled levels). However, the new concerns mentioned above are reflected in a new kind of authority, which allows EPA under certain circumstances to promulgate regulations that will not achieve a 75-percent reduction, at least for some time.

EPA has two options by which it may modify the statutorily prescribed standard. These are contained in Sections 202(a)(3)(B) and (C) and Section 202(a)(3)(E) of the Clean Air Act as amended 1977 (42 USC 7401 et seq.). Briefly, under the former option EPA may depart from the 75-percent reduction only if two tests are met: (1) that the standard cannot be achieved by technology "reasonably expected to be available for such model year without increasing cost or decreasing fuel economy to an excessive and unreasonable degree" and (2) if the National Academy of Sciences has not issued a report with findings "substantially contrary" to those of EPA. Under the authority of Section 202(a)(3)(B) once these facts are determined EPA may revise the 75-percent standard to a level that is "the maximum degree of emission reduction which can be achieved by means reasonably expected to be available for production of such period." This standard will be temporary (for three model years only). Under the second option, EPA may revise the statutory standard after a "pollutant-specific study" of the health effects of the pollutant in question has been conducted. Then, EPA may promulgate a standard, under Section 202(a)(1), (2), and (3), that reflects the

> greatest degree of emission reduction achievable through the application of technology that the Administration determines will be available... giving appropriate consideration to the cost of applying such technology within the period of time available to manufacturers and to noise, energy and safety factors associated with the application of such technology.

It should be noted in this context that in adding these various "escape clauses," Congress specifically said that it was ignorant of the technological feasibility of its 75-percent reduction goal. (U.S. Congress, 1977).

EPA has asserted that this second option comes into play only for "health effects reasons." The first option, in EPA's view, comes into play when there is "a feasibility problem." (U.S. Environmental Protection Agency, 1981a). Since EPA believes that "there is a feasibility problem for [heavy-duty diesels]" it proposes to promulgate a temporary standard under Section 202(a)(3)(B).* Thus, the ultimate standard must reflect "the maximum degree of emission reduction which can be achieved by means reasonably expected to be available.

---
* The Clean Air Act contemplated that the regulatory standard for the 75-percent reduction in NO_x emissions would be in place shortly after the passage of the 1977 amendments. Accordingly, section 202(a)(3)(B) of the Act provided for periodic revisions to be promulgated between June 1, 1980, and December 31,

(footnote continued on next page)

REFERENCES

California Air Resources Board. 1976. "Public Hearing on Proposed Changes to Regulations Regarding Exhaust Emissions Standards and Test Procedures for 1979 and Subsequent Model Year Heavy-Duty Engines." Sacramento, Calif.: California Air Resources Board. (Staff report 76-20-2) October 5.

Federal Court of Appeals for the 2d Circuit. 1975. Society of the Plastics Industry v. OSHA, 509 F.2d 1303 (2d Cir. 1975).

Jambekar, A. B., and J. H. Johnson. 1981. "Effect of Truck Dieselization on Fuel Usage." Warrendale, Pa.: Society of Automotive Engineers. (SAE Paper No. 810022.)

U.S. Congress. 1977. Legislative History to Public Law 95-95, U.S. Code Cong. and Admin. News, 95th Cong. 1st sess., p. 1544.

U.S. Environmental Protection Agency. 1980a. "Draft Regulatory Analysis, Environmental Impact Statement and NO_x Pollutant Specific Study for Proposed Gaseous Emission Regulations for 1985 and Later Model Year Light-Duty Trucks and 1986 and Later Model Year Heavy-Duty Engines." Office of Mobile Source Air Pollution Control. Washington, D.C.: U.S. Environmental Protection Agency, November 5.

_____. 1980b. "Control of Air Pollution From New Motor Vehicles and Motor Vehicle Engines: Gaseous Emission Regulations for 1984 and Later Model Year Heavy-Duty Engines." (Final rule.) Federal Register 45(14):4136. January 21.

_____. 1981a. "Control of Air Pollution from New Motor Vehicle Engines: Gaseous Emission Regulations for 1985 and Later Model Year Light-Duty Trucks and 1986 and Later Model Year Heavy-Duty Engines." (Advance notice of proposed rulemaking.) Federal Register 46(12):5838. January 19.

_____. 1981b. "Control of Air Pollution From New Motor Vehicles and New Motor Vehicle Engines: Certification and Test Procedures." (Notice of intent.) Federal Register 46(70):21628. April 13.

_____. 1981c. "Control of Air Pollution From New Motor Vehicles and New Motor Vehicle Engines: Particulate Regulation for Heavy-Duty Diesel Engines." (Proposed rule.) Federal Register 46(4):1910. January 7.

1980, or between June 1 and December 31 of each third year thereafter. EPA has never promulgated regulations to implement the 75-percent reduction. Although the December 31, 1980, date has already passed, EPA believes that the authority of 202(a)(3)(B) can still be used to promulgate a standard with a reduction of less than 75 percent. Were this incorrect, EPA's options (before 1983) would be to revise the standard under 202(a)(3)(E), i.e., for "health effect reasons," or promulgate a 75-percent reduction standard. This issue has some importance, because the factors to be taken into account are different in each circumstance, and the NAS role is nonexistent under 202(a)(3)(E).

Chapter 2
HEAVY-DUTY DIESEL ENGINES

DESCRIPTION OF THE PROBLEM

This introduction discusses the diesel combustion process from both fundamental and empirical points of view, and then summarizes the various proposed NO_x control techniques for diesel engines. The object is to provide the reader with the background necessary to an understanding of the empirical data presented later in the chapter. The material in this section is drawn largely from a review by Henein (1976) and recent compendia of papers on diesel combustion (Society of Automotive Engineers, 1980, 1981).

Diesel Combustion and Emissions

The combustion process is the critical factor in determining the performance, emissions and fuel economy of diesel engines. The diesel combustion process is fundamentally different from that of gasoline engines. In the latter, a premixed charge of air and fuel is ignited by a spark. In the diesel engine, the intake charge consists of air only. This air is compressed to a high temperature (higher than that of the gasoline engine due to the higher compression ratio of the diesel engine). Near the end of the compression stroke a spray of liquid fuel is injected into the engine; the temperature of the compressed air causes the fuel to evaporate, ignite, and burn. (Some new "stratified-charge" gasoline engines use combustion processes similar to those of diesel engines.)

The amount of fuel that is added at the end of a diesel's compression stroke is less than the amount that could be burned completely by the air present. Thus, on an overall but not necessarily a local basis, combustion takes place under lean conditions, and carbon monoxide is readily oxidized to carbon dioxide; carbon monoxide emissions are therefore not a significant problem in diesel engines.

Initially, some of the injected fuel evaporates and undergoes a series of chemical reactions leading to ignition. The time between the start of injection of the fuel and the start of ignition is called the ignition delay. The length of this delay is governed by the rates of the pre-ignition chemical reactions, as well as by the physical process of atomization and vaporization. All of these in turn depend on cylinder temperature and pressure. Actual ignition occurs around the periphery of the incoming fuel jet. In this region, the air-fuel ratio is probably near the stoichiometric ratio (at which the air/fuel ratio could just give complete combustion). After ignition, expansion of the burning mixture and the already present air motion provide the mixing and combustion for the rest of the fuel.

The energy released by combustion heats any unburned fuel, which then undergoes chemical reactions in an oxygen-deficient atmosphere; these reactions are the start of the formation of particulate matter. Although there is some uncertainty about the exact mechanism of particulate formation in diesel engines, it is generally agreed that the chemical reactions occurring in this oxygen-deficient region form species that can react to form large molecules. Nucleation processes can then form particulate matter from these molecules. Particulate matter so formed can then agglomerate. Once formed in the fuel-rich regions of the engine cylinder, the particulates can and do undergo oxidation reactions in the air-rich regions. The extent of this oxidation is limited by residence time and the temperatures and concentrations of oxidizing species present. The particulate matter finally appearing in the exhaust contains carbonaceous material (soot) and adsorbed liquid hydrocarbons.

High temperatures and oxygen, which are conducive to the oxidation of particulates, are also conducive to the formation of oxides of nitrogen. Most of the nitrogen oxides formed in combustion come from oxidation of atmospheric nitrogen, with a much smaller amount from oxidation of fuel-bound nitrogen. The mechanism for oxidation of atmospheric nitrogen, the Zeldovitch mechanism, is very temperature-sensitive; near-equilibrium amounts of NO may be formed at the high temperatures of engine combustion, but very little decomposition of NO will occur after the temperature is rapidly quenched in the expansion stroke of the engine.

Exhaust hydrocarbons consist of unburned and partially burned hydrocarbons. The presence of these hydrocarbons is due to quenching of the combustion reactions, both at the walls of the combustion chamber and in the lean regions of the combustion gases (where the lower temperatures favor mechanisms that lead to partially oxidized hydrocarbons rather than the products of complete combustion).

During diesel combustion pressures and temperatures in the combustion cylinder can reach 2,000 pounds per square inch (13.8 MPa) and 4500°F (2750 K). The complete injection and combustion process requires from 4 to 10 milliseconds. Liquid fuel is injected at pressures ranging from 3,000 to 25,000 pounds per square inch (20.7 to 172 MPa), and the temperatures of the exhaust gases are typically less than 1000°F (800 K).

The combustion process involves the interaction of chemical and physical rate processes. Ignition of the fuel requires good mixing of the incoming fuel jet with the air. The two basic diesel types, direct- and indirect-injection engines, are differentiated according to the ways they accomplish this mixing.

A direct-injection engine has a single combustion chamber (the top of the cylinder plus any cavity in the piston). Air is introduced into this type of engine with some amount of swirl, provided by the inlet system, to aid mixing. Because this motion is fairly small, the fuel system must provide most of the motion necessary for adequate mixing. Consequently, direct-injection engines use high injection pressures, plus multiple small holes in the injection nozzles.

An indirect-injection diesel engine has, in addition to the main combustion chamber, a prechamber into which the fuel is injected, typically through a single-hole nozzle. Mixing is provided by two mechanisms. The first is the air motion produced in the prechamber by the rising piston. The second is the motion of the prechamber gases as they are ignited and expand into the main chamber.

Most heavy-duty diesels are direct-injection engines; most light-duty diesels are indirect-injection engines. However, direct-injection engines tend to have 10 to 15 percent better fuel economy, and manufacturers have been moving toward direct-injection engines for all applications.

Exhaust treatment of gaseous diesel emissions is difficult because of the low temperatures and large oxygen concentrations. Because the three-way catalyst that is used in gasoline engines depends on having very small amounts of oxygen in the exhaust gases, it cannot be used to remove NO_x from diesel exhaust. One possible method for removing oxides of nitrogen from the exhaust is the addition of a suitable reducing agent such as ammonia or hydrogen. Because of the low temperatures of diesel exhaust, it is not possible to do this without the use of a catalyst.

Diesel Engine Trade-offs

Experimental data establish conclusively that there are trade-offs between performance and emission control in diesel engines. The reasons for these trade-offs are known only in a qualitative fashion, as described above. However, it is well known that they do exist and that they are relatively independent of control technique. Most trade-off curves are approximately hyperbolic in shape, so that the first increment of control produces only small degradation of performance while later increments cause accelerating degradation of performance.

Well-recognized examples of such trade-offs are those between NO_x and fuel consumption, particulates and NO_x, and hydrocarbons and NO_x. However, there are other, less well established ones. For example, the use of exhaust gas recirculation (EGR) to control the formation of oxides of nitrogen is widely suspected of degrading durability and driveability, especially in turbocharged diesel engines. There is evidence to suggest that the particulates and other combustion products in EGR gases may adversely influence durability. However, the evidence is not adequate to establish reasonably valid trade-off curves.

Most heavy-duty diesel engines are turbocharged, and turbocharged engines have a response-time problem during transient operation because of turbocharger inertia. Figure 1 presents data illustrating the driveability problem without EGR; it is worse with EGR. As can be seen in figure 1, when fuel flow is increased instantaneously (solid curve, "fuel pump rack position") inertia prevents immediate turbocharger response. This results in inadequate air flow and excessive smoke (solid curve, "smoke opacity"). If the fuel flow is forced to increase more slowly in spite of the driver's desire for immediate power, smoke can be controlled (dotted curve), but at the expense of driveability, since the engine reaches full speed several seconds later than it otherwise would. This problem is compounded during rapid vehicle accelerations, when part of the air entering the turbocharger is replaced by recycled exhaust. One could cut off EGR during acceleration, of course, but this would reduce the control of NO_x emissions.

CONTROL TECHNIQUES

Retarding Injection Timing

Retarding the timing of the injection of fuel is a universally used means of reducing NO_x emissions. Note, however, that fuel consumption and particulate and hydrocarbon emissions tend to increase with injection retard. Increased injection pressures may mitigate these effects, depending on existing levels of injection pressure and emissions. Most injection systems currently vary injection timing with engine load and/or speed. Much work is being done on electronic control of injection timing, to permit more flexible and precise control. The degree of improvement to be expected from electronic controls depends on the sophistication of the mechanical controls they would replace. Data presented in a National Research Council (forthcoming) report on light-duty diesel engines show a 10- to 15-percent improvement in fuel economy due to improved injection systems. The committee contacted technical experts involved in developing these systems, who reported improvements in this range when the latest generation of mechanical controls were replaced by electronic systems.

The major advantage of electronic controllers, in terms of emission control, is that they permit control system components to be programmed based on instantaneous sensing of engine variables. This allows control devices to operate fully when they have the most beneficial effect on emissions and the least deleterious effect on other aspects of engine performance. Under less advantageous conditions, the control devices can be less fully used, so that the engine can meet a given emission standard with relatively small impairment of performance.

Varying the Shape of the Rate-of-Injection Curve

The shape of the rate-of-injection curve affects both instantaneous fuel

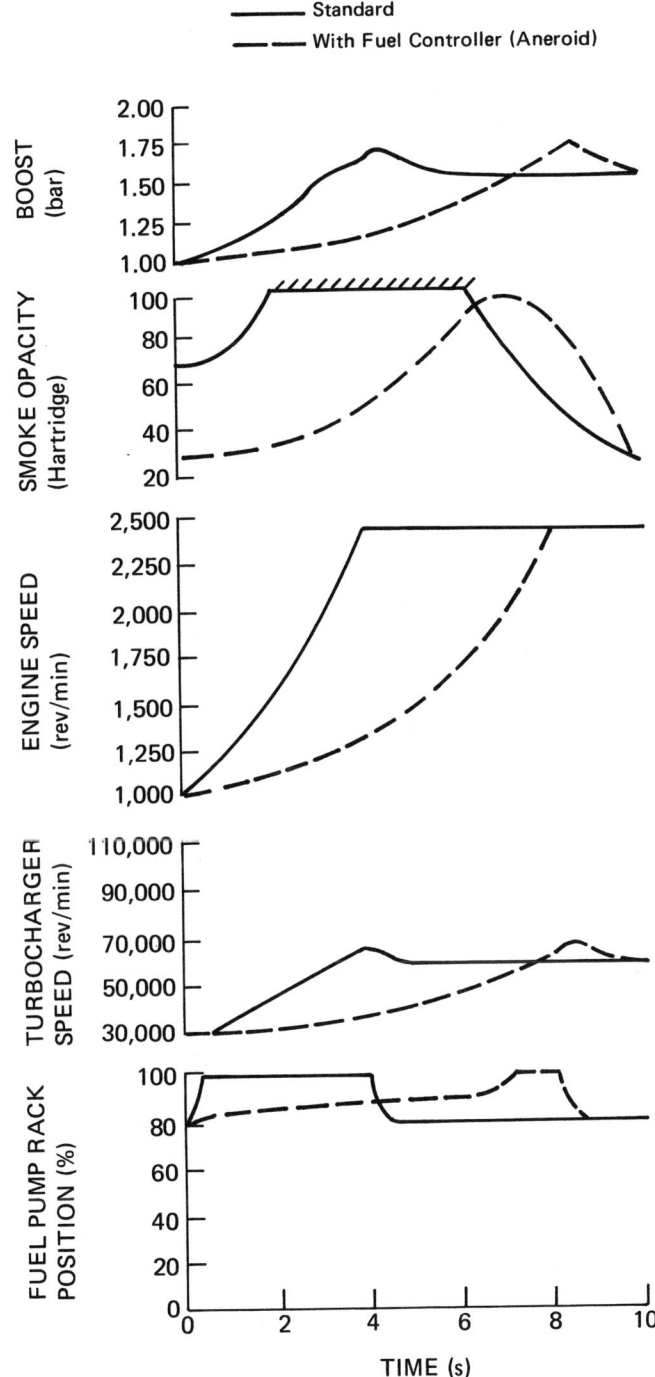

Figure 1 Acceleration of a turbocharged diesel engine against a steady load, showing the trade-off between responses and exhaust smoke (experimental data) (Watson, 1981)

injection pressure (and thus droplet size) and the distribution and mixing of the fuel with the air. Proposed electronic controls for the fuel injection system are intended to control mainly injection timing. However, one can envision systems for varying the shape of the rate-of-injection curve, and laboratory systems to do this have been built. Too little data have been reported, though, to allow an estimate of the benefits in emissions control. Developing a reasonably simple and inexpensive system that can appropriately vary the shape of the rate-of-injection curve during transient operation is a formidable task.

Turbocharging and Charge Cooling

Almost all new heavy-duty diesel engines are turbocharged. Thus, any reduction in emissions due to turbocharging must come from cooling the charge air rather than from turbocharging itself. In general, reducing charge temperature decreases NO_x formation by reducing combustion temperatures.

Modifying Engine Designs

While no dramatic breakthroughs are expected, some changes in design details can reduce emissions. For example, decreasing the volume of the injection nozzle sac has already substantially decreased hydrocarbon emissions. Higher fuel injection pressures have also been beneficial, although there is some evidence that there may be an optimal injection pressure for a given NO_x emission level, so that further increases in pressures may not monotonically reduce emissions.

Exhaust Gas Recirculation (EGR)

Recirculating exhaust gases through the engine air intake is a well-established NO_x reduction technique in spark-ignition engines. It can also be used in diesel engines. In four-stroke-cycle diesel engines it may also reduce hydrocarbon emissions. However, EGR typically increases emissions of particulates and smoke, slightly increases fuel consumption, and, as indicated in the earlier discussion of trade-offs, decreases driveability and possibly durability. Another problem with EGR is fouling of turbochargers and charge cooling equipment.

Catalytic Controls

The presence of oxygen in the exhaust gases is necessary if an oxidation catalyst is to destroy unburned or partially burned fuel. Since diesel exhaust always contains excess oxygen, diesel oxidation catalysts never suffer from a lack of oxygen. The main concerns about oxidation catalysts have to do with contamination, primarily by particulates, and with overheating. Diesel exhaust temperatures decrease rapidly with load. During light-load operation, the catalyst may become partially plugged; it can then overheat when the load is increased and the particulate material burns.

Reduction catalysts require the addition of some fluid (typically ammonia or methane) to combine with the NO_x in the exhaust. The effectiveness of the process is temperature-sensitive, and thus load-sensitive. The process may cause undue ammonia emissions, and it requires the user to take action (in maintaining supplies of the additive) with no direct benefit in return. There are substantial questions about catalyst durability, and the design of an adequate system is a serious challenge.

Water Injection

Water, injected in emulsion with the diesel fuel, is known to reduce NO_x emissions with little increase (and possibly some decrease) in hydrocarbon and particulate emissions or fuel consumption. There is concern about corrosive effects, storage and use of water under cold conditions, and, again, the requirement of action by the user with no direct benefit in return.

Turbocompounding

Turbocompounding (the placing of a turbine, connected to the crankshaft, in the engine exhaust so that more power may be obtained from the engine) primarily affects engine efficiency and not in-cylinder emissions. Thus, any effect on emissions comes mainly from the production of more work at a fixed yield of pollutants. The main motive for tubocompounding is the reduction of fuel consumption rather than reductions in emissions. Using a thermally insulated engine (see below) or raising turbocharger efficiency increases the attractiveness of turbocompounding.

Insulating Engines Thermally

In an insulated engine, energy rejected from the cylinder exits via the exhaust rather than the cooling system. This extra energy in the exhaust is at a higher temperature than the cooling air. This increases the attractiveness of turbocompounding as well as reducing cooling losses. The use of insulated engines would require development of new engine materials as well as new lubricants. It appears that this technique would not increase, and might slightly decrease, in-cylinder NO_x production, and that it would reduce the production of particulates.

Trapping Particulates

Filtering and/or aerodynamic trapping of particulates in the exhaust is undergoing intensive development for use in light-duty diesel-powered vehicles. The amounts of particulates produced require either frequent cleaning of the trap or some means of oxidizing particulates in the trap. Otherwise the collected particulates would increase engine back-pressure, thus increasing particulate production. Oxidation in the trap would require a tight control of the rate of combustion (probably by controlling the supply of oxygen) so that the trap will not be destroyed by heat. Sophisticated interactive controls, probably designed to control oxidation rates by throttling air entering the engine, would be required.

Using Alternative Fuels

Data from light-duty vehicles suggests that a specially tailored diesel fuel may reduce particulate and gaseous emissions. The same may be true of heavy-duty engines. Methanol and ethanol, which have very low cetane numbers, show particulate emissions significantly lower than those from diesel fuel when burned in diesel engines. If the use of methanol and ethanol becomes widespread, a substantially new engine would have to be developed.

Use of Alternative Engines

In laboratory tests, steady-flow engines, such as gas turbines and Stirling engines, show lower emissions but higher fuel consumption than diesel engines.

1986 AVAILABILITY OF CONTROL TECHNIQUES

NO_x control techniques likely to be available in time to meet 1986 emission standards include variable injection timing and pressure, charge cooling, and possibly exhaust gas recirculation. It is not clear whether electronic control of injection timing will be available by 1986; some systems may be on the market by that date, but whether they will have been available long enough to meet engine development, testing, and production lead times is uncertain. Charge cooling using ambient air will be available, since it requires mainly straightforward hardware changes. The outlook for exhaust gas recirculation is not clear; it should be feasible to recirculate small amounts of exhaust gas in heavy-duty engines without significant losses of driveability and durability, especially if electronic control systems are available. It appears unlikely that substantial amounts of exhaust gas can be recycled satisfactorily in all types of heavy-duty diesel engines.

The System Problem

One must consider the entire engine-fuel system and all of its operating and emissions characteristics when judging the feasibility of meeting a given level of any pollutant. As indicated earlier, one emittant is often reduced at the expense of an increase in another. Thus, it is impossible to address the effect of regulating NO_x emissions without also considering the permitted levels of other emittants. For example, in a given engine, if one fixes the levels of particulate and hydrocarbon emissions and the permitted increase in fuel consumption, the level of NO_x emissions is set within fairly narrow limits. One must consider the levels of all pollutants as well as acceptable changes in fuel economy, durability, and driveability.

AVAILABLE DATA

Most of the emissions data on heavy-duty diesel engines comes from the 13-mode steady-state test cycle. (See Appendix C.) The source of this data is engine certification testing by EPA and the engine manufacturers. More limited emission data are available from the new transient test procedure. The sources of this transient cycle data are EPA and some, though not all, heavy-duty engine manufacturers. Generally accepted correlations between the two test procedures have not yet been developed. Figures 2 and 3 show steady-state emission data for hydrocarbons and NO_x plotted against transient emission data for the same pollutants. (The correlation for carbon monoxide is not especially important for diesel engines, whose carbon monoxide emissions are low.) The NO_x data show an approximate one-to-one correlation between the two test procedures. However, there is considerable scatter, and on average the transient cycle NO_x emissions are about 10 percent lower than those from the steady-state cycle. Furthermore, data for low NO_x levels are not available.

Much more information is available on light-duty diesel engines. However, such data cannot be used directly to obtain quantitative estimates of the emissions from heavy-duty engines. Heavy-duty diesels are generally open-chamber, turbocharged engines; most light-duty diesels are prechamber engines without turbochargers. In addition, the emission test cycles and regulatory definitions of useful life are different.

Essentially all of the data on the effectiveness of the emission control technologies discussed in the previous section were taken using the steady-state test procedure. Such data are useful for engineering analyses of promising control options, but there is no established correlation with results from the new transient test procedure, especially at low NO_x and hydrocarbon emission levels.

With the new transient test procedure, preliminary data suggest significant variation in results from laboratory to laboratory. Table 12 illustrates the magnitude of this problem with data from three different engines tested in two laboratories. While NO_x results differ by about 15 percent from laboratory to laboratory, hydrocarbon and particulate results differ by as much as a factor of two.

NO_x emission standards for heavy-duty diesels have been in effect since 1974. (See Table 7.) Uncontrolled NO_x emissions from heavy-duty engines manufactured before 1974 ranged from about 11 to 20 g/bhp-h. In 1974 combined hydrocarbon and NO_x emissions were reduced below 16 g/bhp-h, and in 1979 they were reduced further to below 10 g/bhp-h nationwide and 7.5 g/bhp-h in California.

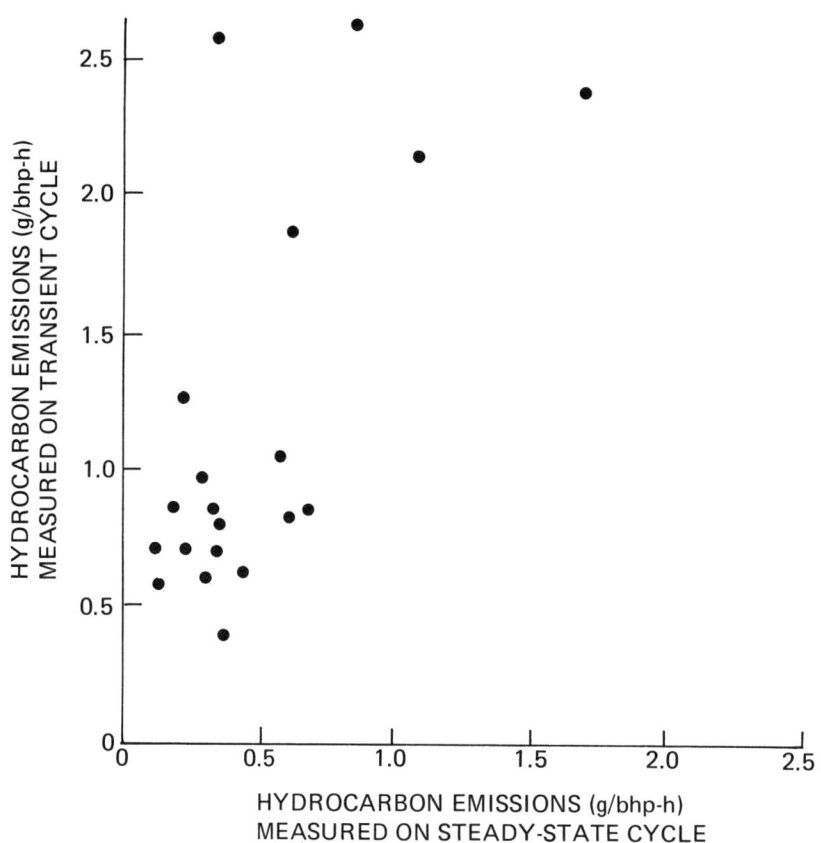

Figure 2   Correlation of hydrocarbon emission data taken with the transient and steady-state test cycles

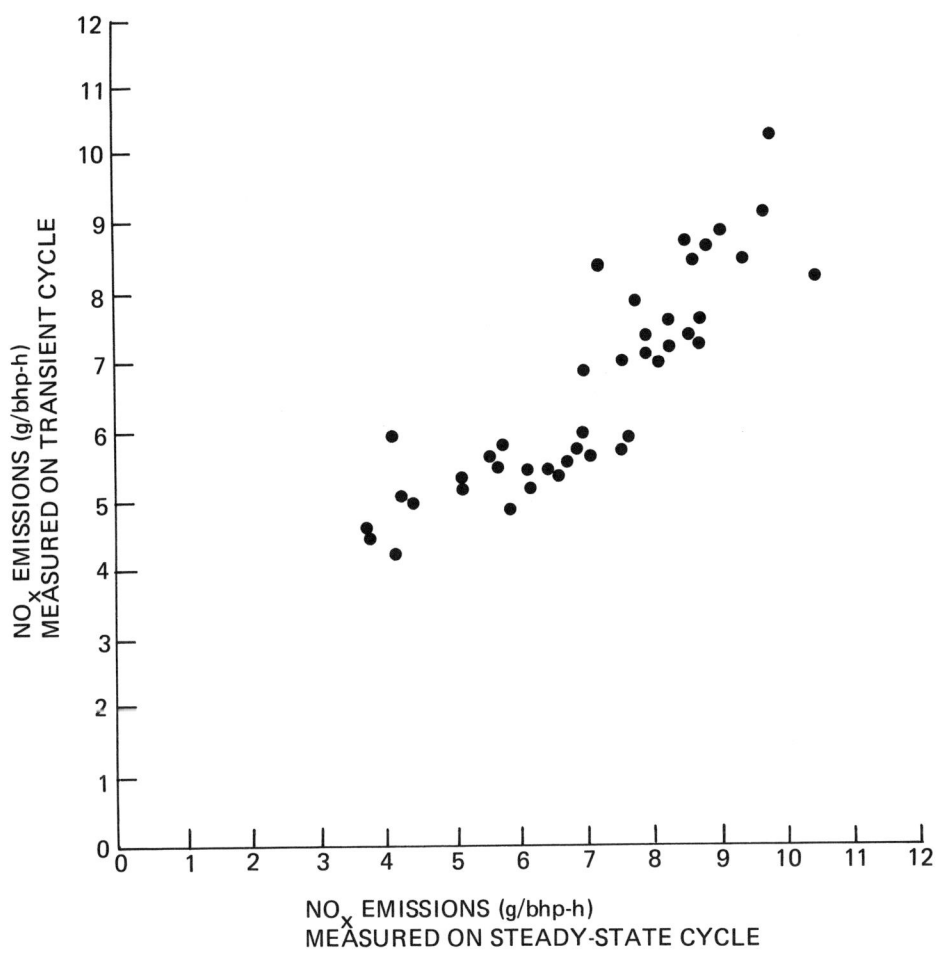

Figure 3 Correlation of NO_x emission data taken with the transient and steady-state test cycles

TABLE 12 Variation in Transient Cycle Emission Test Results

Engine	Test Location	Measured Emission Levels (g/bhp-h)			
		NO_x	Hydro-carbons	Carbon Monoxide	Particulates
1	1	10.0	0.82	5.3	0.78
	2	8.5	0.68	5.0	0.60
2	1	5.7	0.71	2.3	0.55
	2	4.9	0.87	2.6	0.39
3	1	6.2	2.80	2.9	0.98
	2	6.3	1.3	1.4	0.37

SOURCE: Manufacturer's presentation to committee, January 8, 1981.

The 1980 California NO_x standards are 5 g/bhp-h. Table 13 shows 1979 EPA certification data for heavy-duty engines with NO_x emissions levels as measured on the steady-state test. The average over 38 engine families was 7.7 g/bhp-h, with a range from 5.7 to 9.7. Table 14 shows 1980 California NO_x certification data from the steady-state test procedure. The average over 15 diesel engine families from five manufacturers was 4.8 g/bhp-h, with a range from 3.7 to 5.7. Note that 60 percent of the heavy-duty diesel engine families available in the other 49 states are unavailable in California.

The Southwest Research Insitute is conducting a study for the EPA to determine baseline emissions levels of 24 different heavy-duty diesel engines manufactured in model years between 1976 and 1980. Results available to date are given in Table 15. Also, two different measurement techniques and data correction procedures were used for the transient test program. The labeling of some data in Table 15 as "questionable" was done by EPA and Southwest Research Institute. The ranges of these data are 3.8-10.4 g/bhp-h on the steady-state cycle and 4.6-8.7 g/bhp-h on the transient test procedure. The limits on NO_x emissions from current production heavy-duty engines can be summarized as follows: In the 49 states outside of California, emissions range from 6 to 10 g/bhp-h. In California, the range is 4 to 6 g/bhp-h.

The fuel economy penalty of the stricter NO_x emissions requirements in California is probably 3-6 percent. These NO_x emissions levels from current production engines compare with uncontrolled diesel engine NO_x emissions of 11-20 g/bhp-h; that is, Federal standards have cut NO_x emissions by about 20-50 percent, and California standards by about 40-75 percent. Uncontrolled NO_x emissions from heavy-duty diesel engines were one and one-half to three times the uncontrolled gasoline engine emissions baseline of 6.8 g/bhp-h recently established by EPA. (See Chapter 3.) Current controlled diesel emissions are comparable to this uncontrolled gasoline engine baseline level. Note that these summary numbers represent typical values and that the range in values is wide. Different diesel engine types, and different engine designs, can have significantly different NO_x emission levels.

ANALYSIS OF DIESEL NO_x EMISSION CONTROL DATA

A wide variety of diesel engines is produced for the heavy-duty vehicle market. Five domestic manufacturers, with 39 engine families, share over 90 percent of this market. Three-fourths of these engines are turbocharged, and 40 percent of the turbocharged engines are aftercooled. Because the available data on NO_x emission control technology are limited to only a few of these many engine types and designs, it is possible only in general terms to discuss the technology for reducing NO_x emissions. An approximate quantitative assessment of the technology is possible, but the limitations of the data base and the diversity of diesel engines in production must be borne in mind. Different control approaches are unlikely to work equally well on all diesel engines.

All manufacturers expect to use turbochargers on almost all engines, to use intercooling and/or aftercooling, and to use improved fuel injection

TABLE 13 1979 Diesel Engine Family Certification Data

Manufacturer	Engine, Engine Family	Engine Cycle	Turbo-charger	Inter-cooled	After-cooled	Injection Timing	Compression Ratio	CID	Rated Speed	Rated bhp	ECS[a]	1979 Certification 13-Mode NO_x	Southwest Research Institute 13-Mode NO_x	Southwest Research Institute Transient NO_x[b]
GM	4L-53T	2	X			8°,10°	18.7	212	2500	155-170	FM	7.2		
GM	6L-71N	2				13°,15°	18.7	426	2300	184-239	--	8.2		
GM	8V-71N	2				12°,13°	18.7	568	2300	248-316	--	7.7		
GM	6V-71NC	2				10°,13°	18.7	426	2100	160-190	TD, SPL	8.2		
GM	8V-71NC	2				12°,13°	18.7	568	2100	230-270	TD, SPL	7.2		
GM	6V-92TA	2	X		X	10°,14°	17.0	552	2100	300-350	TD, SPL	7.9	7.58	5.83
GM	8V-71TA	2	X		X	12°,14°	17.0	568	2100	350	TD, SPL	6.5		
GM	8V-92TA	2	X		X	11°,13°	17.0	736	2100	435	TD, SPL	7.0		
GM	6L-71T	2	X			11°,14°	17.0	426	2100	260-270	TD, SPL	9.4		
CEC	091 (NH 230,250)	4				19°	15.8	855	2100	220-240	--	7.9		
CEC	092A	4				19°	15.0	855	1900	293	--	8.7		
CEC	092C	4	X								AFC, SP			
CEC	093E (NTC 350,400)	4	X		X	19°	14.3	855	2100	400	SPL, AF	8.7	5.76, 7.81	4.91(5)(E);6.98(7.44)
CEC	172A (VTB 903,350)	4	X					903		350	SPL	7.8		
CEC	172C (" ")	4	X			21°	16.6	903	2100	275	AFC, SP	7.5		
CEC	192B (NT 450)	4	X			18.5°	15.5	1150	2100	450	AFC, SP	8.6	6.93	5.74(6.33)
CEC	193 (MTB 600)	4	X		X	18.5°	14.5	1150	2100	600	AFC, SP	9.7		
CEC	221 (V555)	4				22°	17.0	555	3300	216	--	8.7		
CEC	222 (VT 225)	4	X			16.2°	16.2	555	3000	225	--	8.3		
IHC	DT-466	4	X			16°	16.3	466	24-2600	210	FM, SPL	8.4	5.96	5.67(5.90)
IHC	9.0-Liter	4		X		16°	19.1	551	2800	180	PCV	6.4		
IHC	DTI 466B	4	X			15°	16.3	466	2600	210	FM, SPL	6.4	5.69	5.56(6.67)(E)
Mack	8 (ETZ 1005)	4	X			18°	15.0	998	2100	354	SPL	8.7		
Mack	9 (ENDT 676)	4	X		X	19°	14.99	672	18-2100	283-315	SPL	7.9	6.11	5.25 (6.73)
Mack	10 (ETAZ(B)1005A)	4	X		X	17°	17.0	998	2100	392	SPL	7.9		
Mack	11 (ETZ 675)	4	X			18°	17.0	672	2100	235	SPL	8.2		
Mack	S1B (ETZ 4T7B)	4	X			15°	15.5	475	2400	210	SPL	8.1		
Cat	3 (3208)	4				16°	16.5	676	2800	160-210	--	8.3		
Cat	4 (3306)	4				12°	17.5	638	2200	250	FRC, SPL	5.7		
Cat	9 (3406)	4	X			10°	16.5	893	2100	325	AFRC, SP	6.8		
Cat	10 (3406)	4	X		X	10°	16.5	893	2100	375	AFRC, SP	5.7	5.13	5.12 (5.41)
Cat	11 (3406)	4	X			28°	14.5	893	2100	300-325	AFRC, SP	9.0		
Cat	12 (3408)	4	X		X	11°	15.3	1099	2100	450	AFRC, SP	5.8		
Cat	13 (3208)	4				16°	16.5	636	2800	200	--	6.3		
Cat	14 (3306)	4			X	8.5°	17.5	638	2200	245	EGR	4.8		
Cat	15 (3408)	4	X			28°	14.5	1099	2100	400	AFRC, SP	7.4		
Cat	16 (3406)	4	X		X	26.5°	14.5	893	19-2100	350-380	AFRC, SP	8.7	7.20	7.41 (8.41)
Cat	17 (3408)	4	X		X	28°	14.5	1099	2100	450	AFRC, SP	6.6		
												(7.66)[c]	6.46[c]	5.83 (6.64)[c]

[a] EGR = exhaust gas recirculation; FM = fuel modulator; TD = throttle delay; AFC = air/fuel ratio control; PCV = positive crankcase ventilation; SPL = smoke puff limiter; ECS = emission control system.

[b] CVS bag-sampled NO_x and, in parentheses, dilute continuous sample. Where one of the two sampling techniques was not used, an approximation is given based upon the observed average (1.2) ratio of the two methods, and is indicated by an (E).

[c] Average.

SOURCE: U.S. Environmental Protection Agency, 1980.

TABLE 14 1980 NO_x Levels for Diesel Engines Certified for Sale in California (May 1980)

Manufacturer	Engine Family	NO_x Control System	1980 Certification NO_x[a]	Transient NO_x[b]
Caterpillar	12	–	4.7	–
	13	EGR	4.8	–
	14	–	5.2	–
	16-C	–	4.1	–
Cummins	093G	–	3.7	–
	092	–	4.7	–
Detroit Diesel	8V-92TA	–	4.9	–
	6L-71TA	–	5.2	–
	8V-8.2	–	5.0	–
	6V-92TA	–	4.9	–
International Harvester	DT-466B	–	4.2	–
	DTI-466B	–	4.9	–
Mack	11	–	5.2	–
	10	–	5.7	–
	12	–	4.1	4.61 (5.15)
			(Average, 4.75)	

[a] Obtained from telephone conversation with California Air Resources Board.

[b] Bag-sampled and (in parentheses) dilute continuous integrated.

SOURCE: U.S. Environmental Protection Agency (1980).

TABLE 15 Southwest Research Institute Diesel Baseline Emissions Summary as of June 1, 1980

	Engine Number	Values Given by Transient Test Cycle (g/bhp-h)					Values Given by Steady-State Test Cycle (g/bhp-h)			
		Hydro-carbons	Carbon Monoxide	NO_x^a	Partic-ulates[b]	Number of Tests	Hydro-carbons	Carbon Monoxide	NO_x	Number of Tests
1.	1978 Caterpillar 3208	3.38	3.80	(5.84)	0.79	7	1.70	3.41	4.03[c]	4
2.	1976 Cummins NTC-350	0.68	4.99	(8.51)	0.60	4	0.24	2.19	9.36[c]	2
3.	1978 DDA 6V-92T	0.78	3.15	(7.12)	0.54	2	0.56	2.54	8.61[c]	2
4.	1979 Cummins NTCC-350	0.87	2.59	(4.91)	0.39	2	0.32	3.30	5.76[c]	2
5.	#1 Fuel #2 Fuel	1.52 1.30	3.77 4.35	(5.33) (5.69)	0.69 0.79	2 2	0.84 0.69	6.43 8.22	7.09 7.03	2 2
6.	#1 Fuel #2 Fuel	1.16 1.07	1.49 1.79	(5.82) (5.83)	0.48 0.55	2 2	0.98 0.89	1.73 1.95	7.28 7.58	2 2
7.	1979 IHC DTI 466B (California configuration)	0.83	1.48	(5.56)	0.36	2	0.60	1.22	5.69	2
8.	1979 Deutz F5L-912	[Testing stopped due to defective engine]					2.00	6.60	7.28	2
9.	1979 Mack ETAZ(B) 673A	0.80	6.76	6.73(5.25)	0.58	2	0.56	1.30	6.11	2

Engine Number	Hydro-carbons	Carbon Monoxide	NO_x [a]	Partic-ulates [b]	Number of Tests	Hydro-carbons	Carbon Monoxide	NO_x	Number of Tests
10. 1980 Mack ETSX-676	0.39	1.63	5.15(4.61)	0.63	2	0.35	1.83	3.76	2
11. 1979 Cummins VTB-903 [5/]									
#2 Fuel	1.34	1.38	6.33(5.66)	0.37	2	0.71	1.64	6.93	2
#1 Fuel	1.66	1.63	Void(5.58)	0.31	2	0.96	1.82	7.18	2
12. 1979 CAT 3406 (Family 16)	0.60	3.21	8.41(7.41)	0.52	2	0.29	3.02	7.20	2
13. 1979 CAT 340 (Family 10)	0.48	2.02	5.40(5.12)	0.37	2	0.17	1.66	5.20	2
14. 1979 Cummins "Big Cam" NTC 350	0.62	1.60	7.43(6.98)	0.40	2	0.38	2.62	7.81	2
15. 1979 IHC DT-466	0.84	2.07	5.90(5.67)	0.53	2	0.66	2.47	5.95	2
16. 1979 DDA 6V-92TA[e] (Large injectors)	0.76	2.83	8.69	0.54	2	0.68	2.92	8.78	2
17. 1979 DDA 8V-71TA[e]	0.58	2.38	7.32	0.38	2	0.64	3.10	8.24	2
18. 1979 Cummins NTC-390 [e/]	0.81	2.41	8.28	0.59	2	0.34	2.70	10.40	2
19. 1979 Cummins NH-250	------Currently being tested------								

TABLE 15 (cont.)

	Values Given by Transient Test Cycle (g/bhp-h)				Values Given by Steady-State Test Cycle (g/bhp)			
Engine Number	Hydro-carbons	Carbon Monoxide	NO_x[a]	Partic-ulates[b]	Number of Tests	Hydro-carbons	Carbon Monoxide	NO_x	Number of Tests
20. 1979 DDA V8-8.2									
21. 1979 DDA 6L-71T									
22. 1974 Caterpillar 208 (Family 13)			Awaiting testing						
23. 1979 Caterpillar 208 (Family 3)									
24. 1979 Deutz F5L-912									

[a] All transient NO_x values for Engines 1–8 were determined using bag sampling. After Engine 8, all NO_x values also measured using continuous dilute integrated sampling. All bagged results are enclosed in parentheses. Engine 8, all transient NO_x results were calculated using a humidity correction factor of 1.000. No values for Engines 1–8 were calculated using NO_x correction factor specified in the February 13, 1979, Federal Register (the 1983 heavy-duty hydrocarbon and carbon monoxide notice of proposed rulemaking). Recalculation will occur as time permits.

[b] Calculated per "Draft Recommended Practice for Measurement of Gaseous and Particulate Emissions from Heavy-Duty Engines Under Transient Conditions," EPA Technical Report No. SDSB-79-18, by E. Danielson.

[c] Questionable data.

[d] These data have changed since last reporting due to subsequent discoveries of equipment malfunction. Formal data emissions incorporating the changes will follow.

[e] Tentative data pending formal data submission from Southwest Research Institute.

SOURCE: Informal Communication, U.S. Environmental Protection Agency, Office of Mobile Source Air Pollution Control, Ann Arbor, Mich., December 17, 1980.

systems and more sophisticated timing controls. Many manufacturers anticipate that electronic engine controls will be available in the 1980's. There is considerable debate in the industry about the feasibility of using exhaust gas recirculation (EGR) to reduce NO_x emissions substantially. Very limited data are available on the effect of EGR on all aspects of heavy-duty engine operation. High EGR rates have been shown to reduce NO_x emissions by as much as 80 percent, but also to result in much higher particulate and hydrocarbon emissions and higher fuel consumption. Lower EGR rates, reducing NO_x emissions by perhaps 50 percent, have been shown to reduce hydrocarbon emissions also, but at the expense of a 50-percent increase in particulate emissions. There is much concern about the effect of EGR on engine durability. The duty-cycle load requirements of heavy-duty diesel engines are much more demanding than those of light-duty diesel engines (in which EGR is expected to be used widely).

Data from U.S. engine manufacturers with various NO_x emissions control systems that are feasible for production by the mid-1980s are summarized in Figures 4-6. These figures show the relations between different levels of NO_x emission control and (respectively) fuel consumption, particulate emissions, and hydrocarbon emissions. These trends have been obtained from a relatively small number of individual engines. The data come from engines at low mileage, and they thus do not take into account deterioration in emissions control with increasing use. The upper edge of each band corresponds to simpler emission control systems, generally based on engines of current technology using retarded injection timing as the main control technique. The lower edge of the band corresponds to more sophisticated emission control systems, not yet in production. These include all the techniques listed in this chapter under the heading "1986 Availability of Control Techniques," including exhaust gas recirculation.

Figure 4 shows the fuel consumption changes associated with different levels of NO_x control. The fuel consumption data are for several different engines and control systems. The baseline fuel consumption for each engine and control system is the fuel consumption at current NO_x emissions levels of 8 g/bhp-h. The baseline fuel consumption will be different for different engines. A new engine, of a different design, could have both lower fuel consumption and lower NO_x emissions than another engine. The data, from the 13-mode steady-state test procedure, come from several engine manufacturers and include both four-stroke-cycle and two-stroke-cycle engines. Figure 4 shows that reducing NO_x emission levels below about 8 g/bhp-h results in fuel economy penalties that increase as NO_x levels are lowered. For example, technology under development shows a fuel consumption penalty of about 4 percent at NO_x emissions levels of 5 g/bhp-h, relative to the same technology at NO_x levels of 8 g/bhp-h. Such an engine might have both lower NO_x emissions and lower fuel consumption than an engine of less advanced technology, but data available to this committee show that even such advanced engines suffer increases in fuel consumption when modified to reduce NO_x emissions.

Figure 5 shows that particulate emissions (measured on the 13-mode test cycle) also increase as NO_x emissions are reduced, reaching a level of about 1

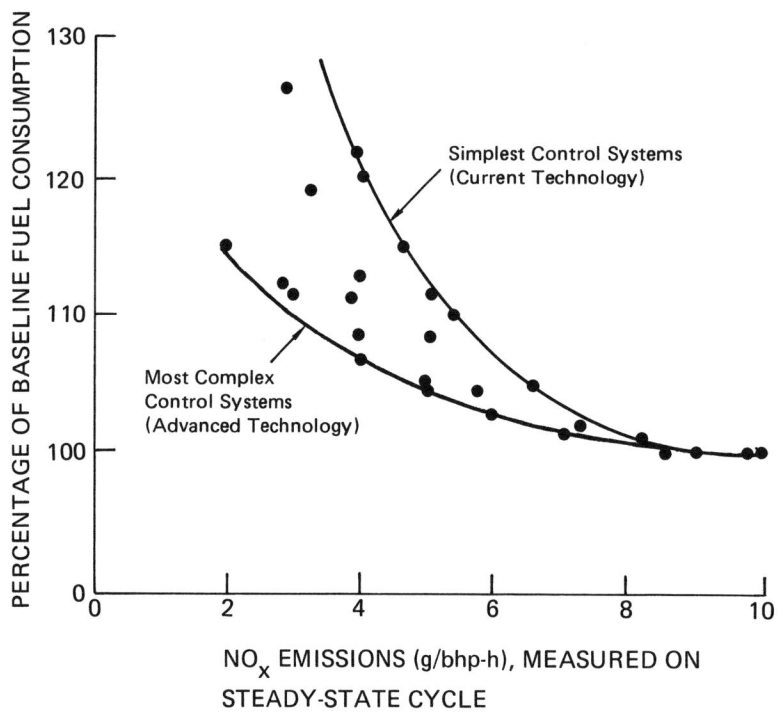

Figure 4 Fuel consumption vs NO_x emission level

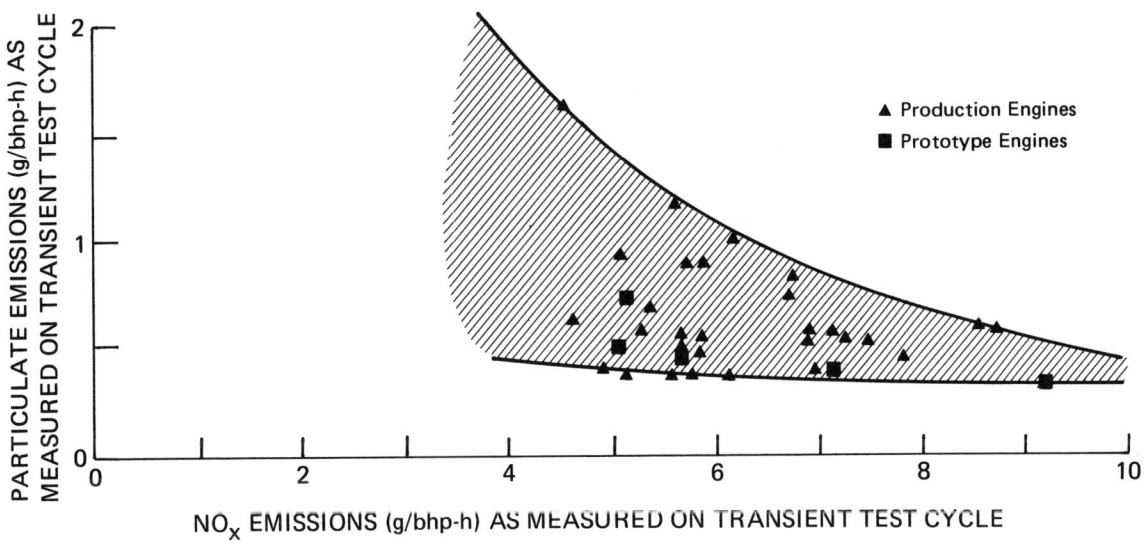

Figure 5   Particulate emission level vs NO_x emission level

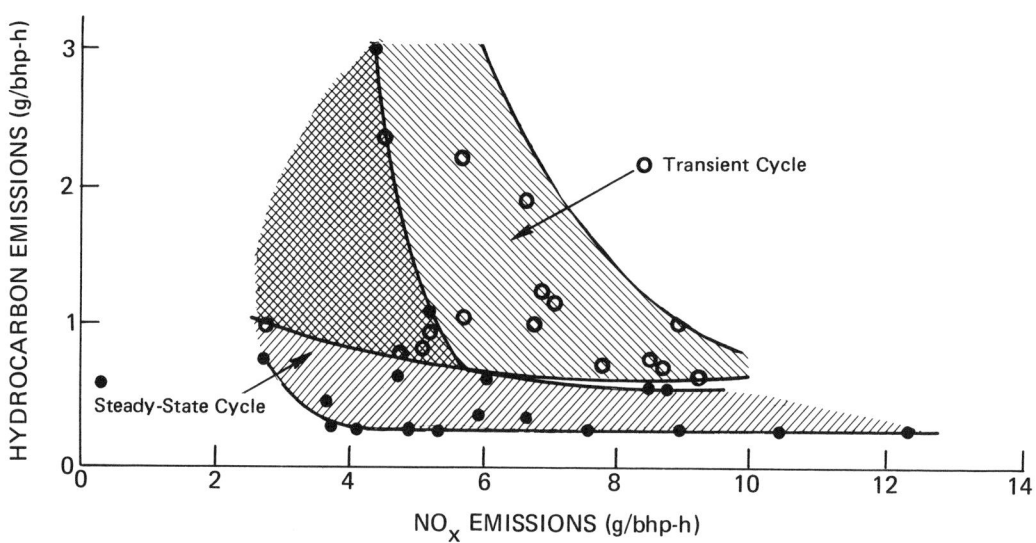

Figure 6 Hydrocarbon emission level vs NOx emission level. Solid circles represent data taken with the transient test cycle, and open circles represent data taken with the steady-state test cycle.

g/bhp-h at NO_x emission levels of 5 g/bhp-h. Particulate emission levels of the current 49-state engine are about 0.6 g/bhp-h. These particulate emissions are engine-out emissions from engines and emission control systems that do not include exhaust particulate traps. Whether particulate traps, which would provide additional reductions in particulate emissions, are feasible for 1986 is unclear. The relationship between NO_x and particulates is described in qualitative terms in the introduction to this chapter. The specific engine data in Figure 5 represent different engines and different operating conditions for the same engines. The quantitative differences indicated in the figure are consistent with the earlier qualitative description.

Figure 6 demonstrates the difficulty of correlating hydrocarbon emission levels measured with the steady-state and transient test procedures; the data from the transient test are higher than those from the steady-state test. Hydrocarbon emissions rise as reductions in NO_x emissions become substantial. The hydrocarbon emission levels of different engines and NO_x emission control systems vary widely. For comparison, the 1984 transient cycle standard for hydrocarbons is 1.3 g/bhp-h.

Two other factors must be discussed before these data can be compared with possible emission standards. These are the Selective Enforcement Audit (SEA) requirements (with a specified Acceptable Quality Limit, or AQL) and the deterioration in emission control over the life of the engine. EPA originally proposed that a 10-percent AQL be used for the SEA of new engines. (See Chapter 1.) A new EPA proposal is a 40-percent AQL. The audit requirement, plus appropriate allowance for deterioration, requires low-mileage emission targets for individual engines to be 20-40 percent below the value of the standard. (For example, low-mileage targets for 1984 model year engines are about 0.8 and 8.5 g/bhp-h for hydrocarbons and NO_x, respectively, in order to meet standards of 1.3 and 10.7 g/bhp-h.) Since data are available on neither engine variability at lower NO_x, hydrocarbon, and particulate levels for advanced control systems nor deterioration of emission control over the useful life of the engine, only rough estimates of low-mileage targets in relation to possible standards can be made.

The above data have been used by the committee in estimating the fuel consumption penalties, and the increases in HC and particulate emissions that result from different levels of NO_x emission controls. Table 16 shows the estimated hydrocarbon and particulate emissions levels and fuel consumption penalties that should be achievable in the mid-1980s at various NO_x emission levels. Only low-mileage targets are given in the table; insufficient information is available to estimate the necessary allowance for deterioration and variability to determine the appropriate standard. The values shown correspond to the lower one-third of the bands shown in Figures 4-6, since only the most promising technology will be used in production. However, it is not reasonable to assume that all production engines can reach the values at the lower edges of the bands, owing to the practical requirement that radically new technology be phased in over a period of several years. Assuming appropriate deterioration factors and margins to accommodate engine-to-engine variability (based on data from current production engines), the standards that could be met would be between 1.2 and 1.4 times the low-mileage values given in Table 16.

SUMMARY

NO_x emission levels from current heavy-duty production diesel engines sold in all states except California are about half the emission levels of uncontrolled diesel engines. Diesels sold in California have NO_x emission levels about one-third the uncontrolled levels.

The uncontrolled diesel NO_x baseline is one and one-half to three times as high as the uncontrolled heavy-duty gasoline engine baseline of 6.8 g/bhp-h. The available data for evaluating the potential of NO_x control technology to achieve levels substantially lower than current levels are limited. Only approximate engineering estimates can be made of this potential.

Trade-offs between NO_x control and hydrocarbon and particulate control, fuel consumption, and driveability are encountered as NO_x standards are made more stringent. Only modest additional NO_x control can be achieved without fuel economy penalties at lower hydrocarbon and particulate emission levels than those currently being achieved.

The more promising emission control techniques being developed show that low-mileage NO_x emission levels of 6 g/bhp-h on the transient test cycle can be achieved with engine-out particulates emissions of 0.5-0.7 g/bhp-h and hydrocarbon emissions of 0.7-1.4 g/bhp-h, and with a fuel consumption penalty of 2.5-4 percent. Low-mileage NO_x levels of 4 g/bhp-h can be achieved with higher particulate and hydrocarbon emission levels and with a fuel consumption penalty of 7-12 percent.

The available data on engine-to-engine variability with the appropriate test procedure, and the data on deterioration in emission control over the engine's useful life, are inadequate to relate NO_x, hydrocarbon, and particulate emission levels of individual engines at low mileage to appropriate levels for the standards. The NO_x standards corresponding to the low-mileage emission levels in the preceding paragraph would be 1.2 to 1.4 times the low-mileage levels.

TABLE 16 Trade-offs Between NO_x Emissions and Particulate Emissions, Hydrocarbon Emissions, and Fuel Consumption (Low-Mileage Emission Levels, as Measured by the Transient Test Procedure)[a]

NO_x	Emissions (g/bhp-h)		Fuel Consumption Penalty (%)
	Particulates	Hydrocarbons	
8	0.4-0.5	0.6-0.8	0
6	0.5-0.7	0.7-1.4	2.5-4
4	0.6-1.0	0.8-1.7	7-12
2	(b)	(b)	15-20

[a] Data on NO_x emissions and fuel consumption are from the steady-state test procedure; for this purpose the two tests are assumed equivalent.

[b] Unknown; too few data are available to permit realistic estimates.

REFERENCES

Henein, N. A. 1976. "Analysis of Pollutant Formation and Fuel Economy and Control in Diesel Engines." *Progress in Energy Combustion Science,* 1:165-207.

National Research Council. Forthcoming. *Diesel Technology*. Diesel Impacts Study Committee, Technology Panel. Washington, D.C.: National Academy Press.

Society of Automotive Engineers. 1980. *Diesel Combustion and Emissions*. Warrendale, Pa. (P-86)

_____. 1981. *Diesel Combustion and Emissions, Part II*. Warrendale, Pa. (SP-484)

U.S. Environmental Protection Agency. 1980. "Draft Regulatory Analysis, Environmental Impact Statement and NO_x Pollutant Specific Study for Proposed Gaseous Emission Regulations for 1985 and Later Model Year Heavy-Duty Engines." Office of Mobile Source Air Pollution Control. Washington, D.C.: U.S. Environmental Protection Agency.

Watson, N. 1981. "Transient Performance Simulation in Analysis of Turbocharged Diesel Engines." Warrendale, Pa.: Society of Automotive Engineers. (SAE Paper No. 810338.)

Chapter 3
HEAVY-DUTY GASOLINE ENGINES

INTRODUCTION

Heavy-duty and light-duty gasoline engines are often similar in design and construction. Some manufacturers, in fact, use the same engines (with modifications) in both heavy- and light-duty vehicles. However, a significant difference between the two applications is the greater fraction of high-power use typical of heavy-duty service.

Because light-duty and heavy-duty engines are often similar, basic emission control techniques are common to the two applications. In the past, emission control requirements for light-duty engines have been more stringent than those for heavy-duty engines of the same model year. For this reason, the control technology used on current light-duty engines is a useful source of information on techniques for meeting future, more stringent regulations on heavy-duty engine emissions.

Control Techniques

Emission controls for gasoline engines are based on engine modifications and/or the addition of exhaust-treatment devices such as catalytic converters. Engine modifications include changes in the air/fuel ratio and spark timing; redesign of combustion chambers, intake manifolds, and fuel metering systems; changes in valve timing; and the addition of exhaust gas recirculation (EGR).

These modifications have been used by themselves to meet moderately stringent emissions standards. However, as standards become more stringent this approach typically results in deteriorating fuel economy and engine performance. As a consequence, catalytic controls are now used in light-duty vehicles to meet the more stringent emission limitations while at the same time allowing the engines themselves to be optimized for fuel economy and performance.

Noncatalytic after-treatment of emissions, by addition of a pump to inject air into the hot exhaust manifold, has been used in the past to reduce the hydrocarbon and carbon monoxide emissions of light-duty vehicles and is currently used on heavy-duty gasoline engines. This type of emission control can be enhanced by redesigning the exhaust manifold to provide longer residence times at high temperature. Some noncatalytic oxidation of hydrocarbons and carbon monoxide may also be used in combination with catalytic control systems.

The use of catalytic converters for emission controls on gasoline-powered light-duty vehicles is virtually universal in the United States. Some vehicles use oxidizing catalysts, which control only hydrocarbon and carbon monoxide emissions, and control NO_x emissions by engine modifications. Most catalytic systems on 1981 light-duty vehicles, however, use three-way catalysts (TWC), which simultaneously control hydrocarbons, carbon monoxide, and NO_x. The operation of these three-way systems requires close control of air/fuel ratios. This is achieved by the use of exhaust gas oxygen sensors and electronic feedback-controlled fuel metering systems. Both carburetors and fuel injection can be used for feedback-controlled fuel metering. The electronic control systems, which are based on microcomputer technology, are capable of monitoring engine variables and ambient conditions such as barometric pressure. The results of these measurements are combined and used to set air/fuel ratio, spark timing, and emissions control system parameters to control emissions while maintaining the best possible performance and fuel economy.

EFFECTS OF ENGINE MODIFICATIONS ON EMISSIONS AND FUEL ECONOMY

The effects of various engine design features on gasoline engine emissions and performance have been studied for several years. General descriptions of these effects may be found in several references (Patterson and Henein, 1972; Obert, 1968; National Research Council, 1974). A brief summary is given under the following heads.

Air/Fuel Ratio

Figure 7 shows the effects of variations in air/fuel ratio on emissions, power, exhaust temperature, and fuel economy. Engines without emission controls (especially heavy-duty engines) are typically calibrated to run rich, to ensure good driveability and high power output. Operation at lean air/fuel ratios, which could provide better fuel economy, has often been avoided to minimize performance and driveability problems.

Spark Timing

Spark timing can be retarded (set to occur closer to the end of the compression stroke) to reduce emissions of hydrocarbons and NO_x. However, this lowers fuel economy and performance.

Exhaust Gas Recirculation

Introducing exhaust gases into the intake charge lowers peak combustion

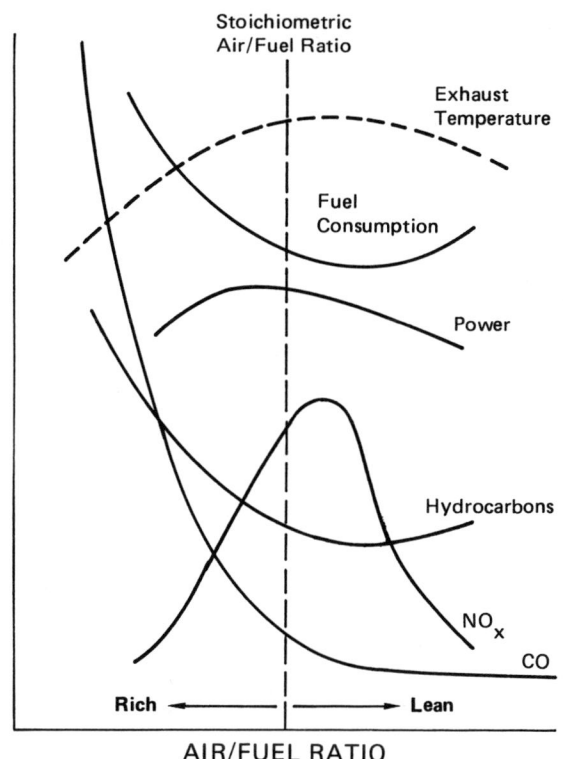

Figure 7 Schematic illustration of the dependence of emissions and other engine parameters on air/fuel ratio. Vertical scale is linear; the curves show the relative variation in the indicated parameters for a typical engine.

temperatures, which in turn yields lower NO_x emissions. Effects on other pollutant species and fuel economy depend on the amount of EGR used and the design of the engine. At low EGR rates, there is little or no effect on these variables. At the high EGR rates required for stringent NO_x control without catalysts, fuel economy and performance can deteriorate significantly. The use of EGR at wide-open throttle settings reduces maximum engine power.

Valve Timing

Modified valve timing, particularly valve overlap (the time interval in which both the intake and the exhaust valve are open simultaneously) provides some internal EGR for NO_x control.

Combustion Chamber Redesign

Reduction of crevices and changes in the surface-to-volume ratio are used to reduce emissions. More sophisticated changes, aimed at better tolerance of EGR, are the subject of current research.

Compression Ratio Changes

Compression ratios are determined generally by the octane rating of the gasoline available. Decreased compression ratios may provide lower emissions of hydrocarbons and NO_x. However, these benefits are usually accompanied by a fuel economy penalty.

Fuel System Modifications

In addition to the overall air/fuel ratio, the cylinder-to-cylinder variance in air/fuel ratio, as influenced by induction system design, is an important consideration. Although the overall air/fuel ratio of an engine may be precisely controlled, the way fuel and air are mixed in the intake manifold can impart to each cylinder a different air/fuel ratio. In engines without emission controls, the overall ratio is usually set rich enough to ensure that the leanest cylinder will have a ratio rich enough to provide adequate performance. Modifications to the intake system that improve the cylinder-to-cylinder uniformity of the air/fuel ratio allow the overall ratio to be set leaner without impairing other engine performance parameters. Such modifications are used to some extent on current heavy-duty engines and to a much greater extent on light-duty engines.

PERFORMANCE OF CURRENT HEAVY-DUTY GASOLINE ENGINES

Correlation of Emission Data
from the Steady-State and Transient Test Cycles

Emissions of current production engines are determined by testing on the nine-mode steady-state test cycle. Some data on the emissions performance of

1979 model year gasoline engines, from both the new transient cycle and the steady-state cycle, are available. These data were obtained by Southwest Research Institute under contract to EPA (California Air Resources Board, 1981). All data were obtained from engines with noncatalytic control systems. Comparisons of emissions for the two test cycles are shown in Figures 8-10. The lines shown on each graph are least-squares linear fits of the data. The corresponding equations and correlation coefficients for the 12 engines tested are given below, for hydrocarbons (HC), carbon monoxide (CO), and NO_x:

$$\text{Transient HC} = 2.26 + 0.553 \times (\text{Steady-State HC})$$

(correlation coefficient, 0.460)

$$\text{Transient CO} = 68.9 + 0.31 \times (\text{Steady-State CO})$$

(correlation coefficient, 0.12)

$$\text{Transient NO}_x = 1.44 + 0.74 \times (\text{Steady-State NO}_x)$$

(correlation coefficient, 0.73).

The poor correlation of the two test cycles for hydrocarbons and carbon monoxide is obvious. The correlation coefficients for these species are not significantly different from zero by a one-tailed t-test at the 0.05 level. The correlation for NO_x is somewhat better, with a correlation coefficient significantly different from zero at the 0.005 level. Nevertheless, data on some engines are far from the correlation line. In addition, none of the data used in obtaining this correlation were at low emissions levels; the lowest NO_x levels were 3.95 g/bhp-h (steady-state) and 4.29 g/bhp-h (transient). These two lowest values were not obtained with the same engine. Also, these data represent relatively new engines that had been run on a dynamometer for 125 hours as certification engines. It is not clear how the correlation between the two test cycles would change if it were run on engines at the ends of their useful lives, at which point the NO_x standard will apply.

Although the above correlation equation for NO_x is an estimate at best, it does give a rough indication of how an engine with a given NO_x emission level on the steady-state cycle will perform on the transient cycle.

Examples of Current-Production Low-NO_x Engines

Table 17 shows data for Chrysler heavy-duty gasoline engines for the 1981 model year. These engines are designed to be used in the lowest range of heavy-duty applications (gross vehicle weight ratings less than 11,000 lb). They are of particular interest because they use oxidation catalysts (with air pumps) to control emissions of hydrocarbons and carbon monoxide. Spark retard and EGR are used for NO_x control. The oxidation catalyst allows the

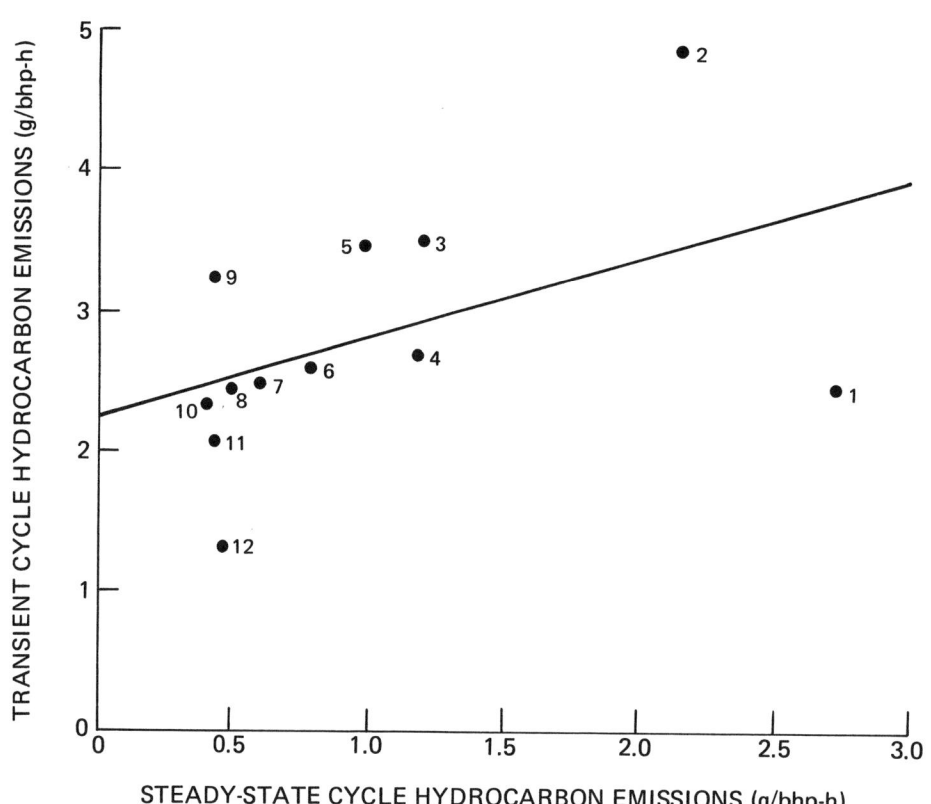

Figure 8 Comparison of hydrocarbon emission data taken with the transient and steady-state test cycles. The numbers by the data points are arbitrary engine identifiers, used also in Figures 9 and 10 (California Air Resources Board, 1981).

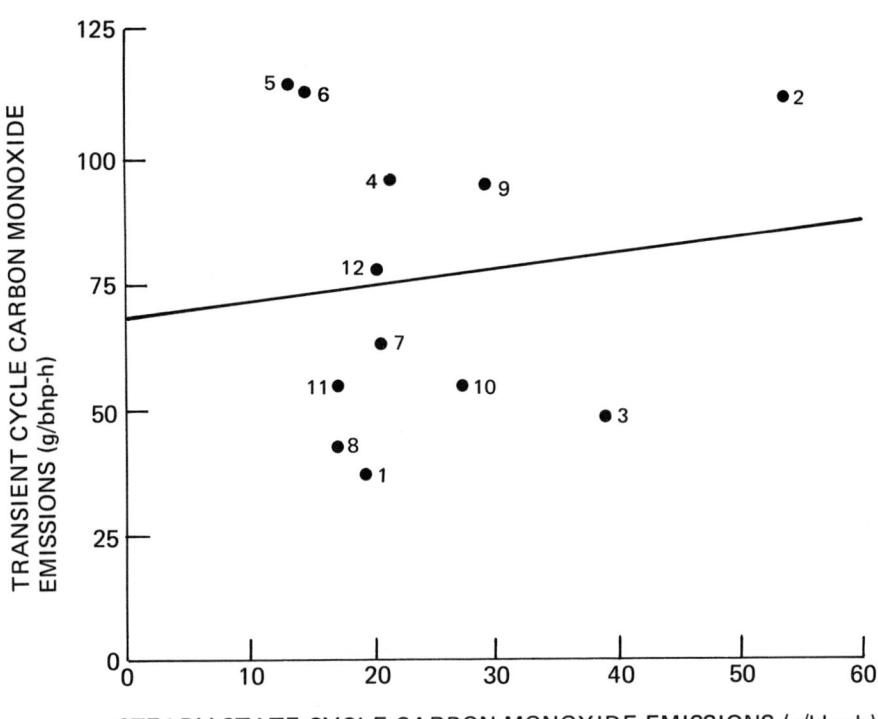

Figure 9   Comparison of carbon monoxide emission data taken with the transient and steady-state test cycles. Numbers by the data points identify the same engines as in Figures 8 and 10 (California Air Resources Board, 1981).

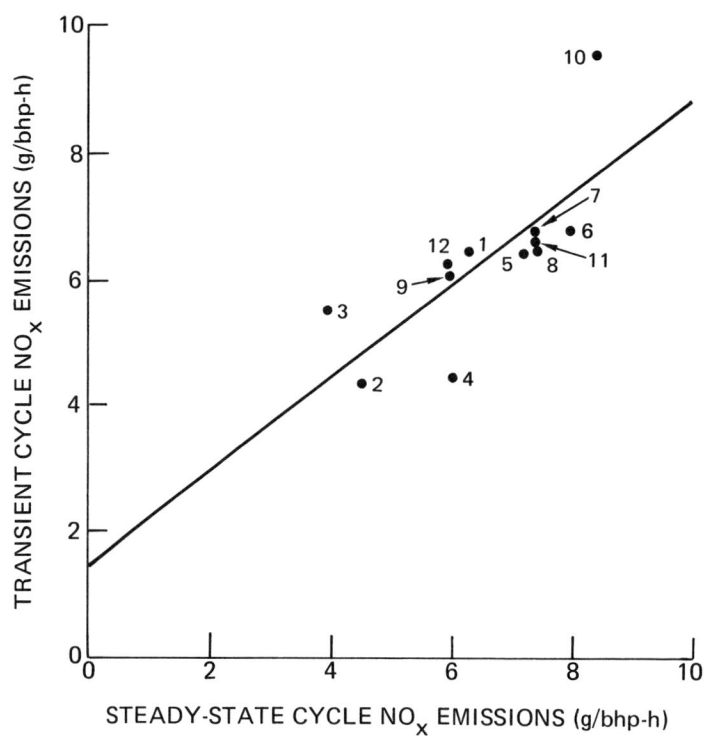

Figure 10 Comparison of NO_x emission data taken with the transient and steady-state test cycles. Numbers by the data points identify the same engines as in Figures 8 and 9 (California Air Resources Board, 1981).

Table 17 Current Emission Levels of Chrysler Heavy-Duty Engines, as Measured by Steady-State Test Procedure

Catalyst	Engine Designation	Emissions (g/bhp-h)[a]			Fuel Consumption g/bhp-h	Maximum Power hp
		Hydrocarbons	Carbon Monoxide	NO_x		
No	360-1	0.40	17.06	5.22	318	190
No	360-3[b]	0.48	8.11	5.13	324	130
Yes	360-1	0.11	10.60	1.78	406	190
Yes	360-1	0.09	3.90	4.04	343	190
Yes	360-3	0.18	14.82	1.87	378	180
Dual	360-3	0.05	11.90	2.32	374	200

[a] Emissions values include deterioration factors measured for 1,500-hour dynamometer run.

[b] The 360-3 engine is designed for more severe use than the 360-1 engine.

SOURCE: Informal communication, Chrysler Corporation, April 1, 1981.

engine to be calibrated rich for better driveability, without sacrificing control of carbon monoxide and hydrocarbons emissions. With catalysts, these engines have very low emissions--significantly lower than the California heavy-duty standards of 1, 25, and 6 g/bhp-h for hydrocarbons, carbon monoxide, and the sum of hydrocarbons and NO_x, respectively.

These very low emissions do carry a significant fuel consumption penalty, however. The catalyst-equipped engines consume 8-17 percent more fuel than their counterparts without catalysts, mainly as a result of the richer air/fuel ratio. Note the accompanying reductions in NO_x emissions of the catalyst-equipped engines.

The data in Table 17 are from the steady-state test cycle. The correlation equation of the previous section, though not validated at these low levels of NO_x emissions, can be used to estimate the NO_x levels these engines would attain on the transient cycle. The lowest steady-state NO_x number(s) in Table 17 then correspond to a value of about 2.8 g/bhp-h on the transient procedure.

Chrysler representatives contacted by the committee were not aware of any durability problems with the oxidation catalysts on their engines. They reported no problems with operation of the catalyst at 90 percent of full load, a required test point on the steady-state cycle.

Another low-emission engine in current production is the International Harvester 345-cubic-inch-displacement engine. This engine had emissions of 0.3, 18, and 2.6 g/bhp-h respectively for hydrocarbons, carbon monoxide, and the sum of hydrocarbons and NO_x, with NO_x emissions of 2.3 g/bhp-h on the steady-state cycle (corresponding to an estimated 3.1 g/bhp-h on the transient cycle, according to the correlation equation). The engine is rated at 162 hp at 3,600 rpm.

The International Harvester engine uses a rich mixture, rather than EGR, to lower NO_x emissions. Two air pumps, one for each exhaust bank, are used to control the carbon monoxide emissions that result from rich operation. International Harvester representatives estimated the highway fuel economy loss in going from the 1979 model year engine to this current version at about 7 percent. The 1979 engine was certified at 5.6 g/bhp-h NO_x emissions on an earlier version of the steady-state cycle. Details of this earlier cycle were modified for the 1980 year model. If the 1979 engine were tested on the latest version of the steady-state cycle, NO_x emissions would probably be higher. Thus, it is not possible to make a direct comparison of the fuel economy changes with specific emissions levels.

The data on the Chrysler and International Harvester engines reported here are not typical of the current mix of production engines. They do, however, indicate the emissions levels that can be achieved with current heavy-duty control technology. Unfortunately, no transient cycle data were available for these engines. Also, it is not known how the engines would perform at the ends of their useful lives, when the proposed new regulations would apply.

EMISSIONS DATA FOR PROTOTYPE 1986 CONTROL SYSTEMS

Emissions test data relevant to the proposed 1986 standards are extremely scarce at present. First of all, the recent implementation of a new heavy-duty transient test procedure involves significant installation expense and effort. In addition, the proposed NO_x standards have only recently been announced.

Limited test data are available for experimental emissions control systems on essentially new engines. However, at present there appears to be no information on the durability of such systems in the laboratory or in use. Since the heavy-duty gasoline vehicle represents a new and potentially severe application of catalytic emission control technology, the durability question is vital to evaluating the feasibility of meeting proposed standards.

Noncatalytic NO_x Controls

Figure 11 represents schematically the systems used for noncatalytic NO_x emissions control. As explained earlier, these can include recalibration of air/fuel ratios, rescheduling of ignition timing, and introduction of EGR. Since the severity of the proposed 1986 NO_x standard virtually dictates the use of catalytic control techology, little information has been developed on the performance of noncatalytic systems at these low NO_x levels. Such data, if available, would be helpful in determining minimum feasible emissions levels for noncatalytic systems and the magnitudes of the fuel economy and performance penalties involved.

While test data are limited, estimates of the potential effectiveness of EGR coupled with changes in engine calibration, extracted from material provided by the General Motors Corporation (1981) are presented in Table 18. These data are based on laboratory engine tests and therefore don't address possible impacts on performance or driveability. Transient test cycle fuel economy values are included, however.

The data in Table 18 were derived from tests run on a prototype 350-cubic-inch-displacement V-8 engine equipped with dual oxidizing converters to meet 1984 hydrocarbon and carbon monoxide standards. Hydrocarbon and carbon monoxide measurements taken with fresh catalyst are well within the standards. With application of EGR, an NO_x emission reduction approaching 50 percent of base engine emissions was realized. The lowest recorded NO_x emission rate was about 3 g/bhp-h. At this emission level, fuel consumption appeared to increase by about 4 percent compared with base engine levels. However, the statistical significance of this increase is not known.

A second estimate of the potential effectiveness of EGR and its impact on fuel consumption can be obtained from data published recently by EPA (Hansel, Cox, and Nugent, 1981). This work was aimed at applying three-way catalyst technology to heavy-duty gasoline engines, but some inferences regarding EGR performance in noncatalytic systems can be drawn. All testing involved a 1978 model International Harvester 404-cubic-inch-displacement V-8 engine

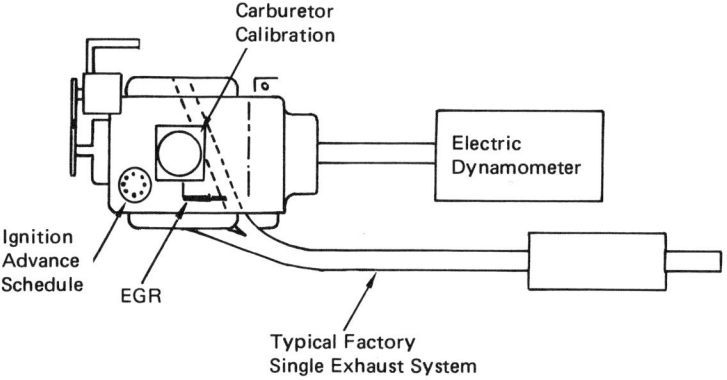

Figure 11 Noncatalytic emission control (Hansel, Cox, and Nugent, 1981)

TABLE 18 Effect of Exhaust Gas Recirculation (EGR) On Exhaust Emissions and Fuel Consumption of 350-Cubic-Inch-Displacement V-8 Engine with Fresh Oxidation Catalyst

Engine Configuration	Emissions and Fuel Consumption, as Measured on Transient Cycle (g/bhp-h)			
	Hydrocarbons	Carbon Monoxide	NO_x	Fuel Consumption
No EGR	0.47	4.01	5.65	245
EGR No. 1	0.55	5.51	3.98	250
EGR No. 2	0.49	5.88	3.40	254
EGR No. 3	0.59	5.04	3.03	259
EGR No. 4	0.67	5.97	2.91	254

SOURCE: Informal communication, General Motors Corporation, February 4, 1981.

equipped with an exhaust oxygen sensor and a prototype feedback-controlled carburetor. In Table 19, data extracted from the same source have been arranged to show the effects of EGR on emissions in the absence of catalytic controls. The EGR system employed is the 1978 production system used in the International Harvester engine, and therefore does not represent a special attempt to approach 1986 standards. The results are nevertheless helpful in assessing EGR effects.

Table 19 suggests that EGR can yield NO_x emission reductions of 30-40 percent with no apparent compromise of hydrocarbon or carbon monoxide emissions. Fuel economy results are mixed, one set of data showing a small improvement and the other showing a small degradation with application of EGR. In both cases, fuel economy changes may be within the range of random variability of the transient test procedure. It is probably reasonable to conclude that the levels of EGR incorporated have no significant effect on fuel consumption.

It should be stressed in summarizing the effectiveness of noncatalytic systems that very little test data defining EGR effects on emissions and fuel consumption using the new test cycle are available. Most important, data for the high EGR rates required to approach the proposed 1986 standard are not available. Based on the General Motors and EPA data presented here, the committee judges that in new vehicles well-engineered systems can achieve NO_x emissions of about 5 g/bhp-h, with little or no fuel economy penalty. EGR combined with engine recalibration can yield NO_x levels of about 3 g/bhp-h, but this may increase fuel consumption by 3-7 percent.

Catalytic Control of NO_x Emissions

Catalytic reduction of NO_x emissions, using three-way catalyst systems similar to those used in late model passenger cars, is being considered for gasoline-powered, heavy-duty vehicles. As shown in Figure 12, these systems can take several forms, involving either single three-way catalyst beds or combinations of three-way catalyst beds and downstream oxidation catalysts to enhance control of hydrocarbon and carbon monoxide emissions.

Data presented in this section will demonstrate that, with fresh catalysts and new engines, three-way catalyst systems are capable of approaching the NO_x emission levels required by the proposed 1986 standards. However, the crucial factor in compliance is catalyst durability over the engine's life as defined by the pertinent regulations. Unfortunately, so far as the committee can determine there are no meaningful data on the durability of heavy-duty three-way catalyst systems. There is strong evidence, however, to indicate that the severity of heavy-duty service may cause catalysts to deteriorate more rapidly than in passenger vehicle service. For this reason, emissions deterioration factors obtained from light-duty vehicles are not applicable to heavy-duty vehicles.

TABLE 19 Effect of Exhaust Gas Recirculation on Exhaust Emissions and Fuel Consumption of a 404-Cubic-Inch-Displacement V-8 Engine[a]

EGR	Emissions as Measured on Transient Cycle (g/bhp-h)			Percentage Reduction in NO_x Emissions	Brake Specific Fuel Consumption (g/bhp-h), With Percentage Change in Parentheses	
	Hydro-carbons	Carbon Monoxide	NO_x			
350-mV Control Point[b]						
No	4.64	43.5	8.04		322	
				(26.7)		(−3.95)
Yes	4.61	39.8	5.89		310	
530-mV Control Point						
No	4.04	49.44	7.88		304	
				(36.2)		(+1.3)
Yes	4.82	49.02	5.03		308	

[a] Engine equipped with exhaust oxygen sensor and feedback carburetor control.

[b] The control point refers to the setting on the electronic feedback carburetor control system, which determines the mean air/fuel ratio. (550 mV = chemically correct; 350 mV = lean.)

SOURCE: Data from Hansel, Cox, and Nugent, 1981.

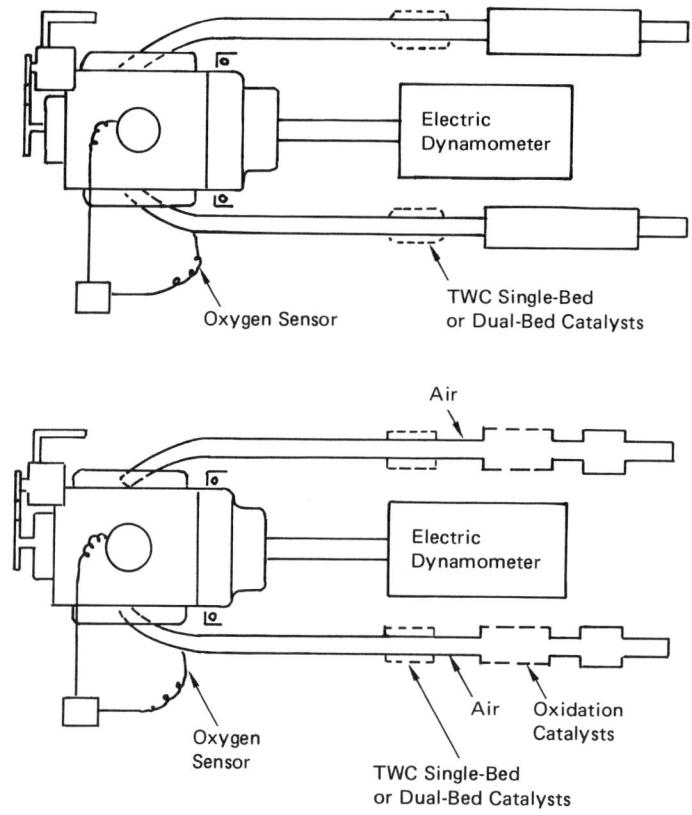
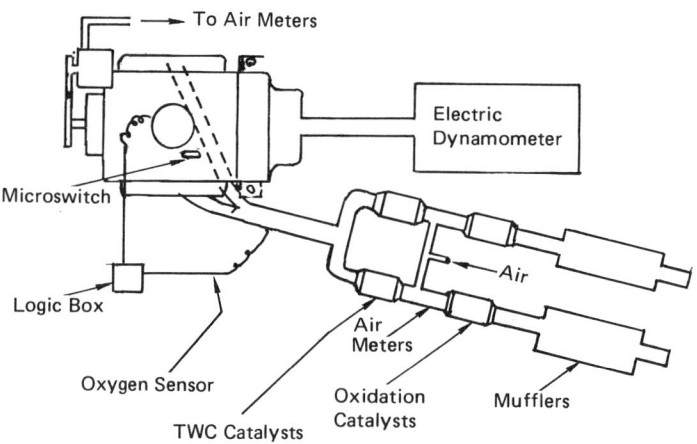

Figure 12 Catalytic NO_x emission control systems (Hansel, Cox, and Nugent, 1981)

New Engine, Fresh Catalyst Emissions Data

Two sets of transient test data suggest that, again with fresh catalyst, three-way catalyst systems can approach or meet the proposed 1986 standards. All of the data indicate a measurable fuel consumption penalty.

Table 20 represents data provided by the General Motors Corporation (1981) for a prototype 1986 system involving a 350-cubic-inch-displacement V-8 engine equipped with dual three-way catalytic converters and an EGR system for comparison purposes, data from a 1984 prototype system using dual oxidizing converters are also tabulated. With the three-way system, NO_x emissions were less than 2 g/bhp-h, while hydrocarbon and carbon monoxide levels were maintained below 1984 limits. Fuel consumption increased by about 5 percent compared with that of the 1984 prototype with no EGR, but was comparable to that of the 1984 engine when EGR was added to bring NO_x emissions to about 3 g/bhp-h. It should be stressed that these results reflect the performance of fresh catalyst materials with less than 10 hours of aging.

Data obtained by the EPA (Hansel, Cox, and Nugent, 1981) for a prototype three-way catalyst system are shown in Table 21. These tests employed an International Harvester 404-cubic-inch-displacement V-8 engine equipped with an exhaust oxygen sensor and experimental feedback-controlled carburetor. Both single three-way catalyst systems and dual-bed systems combining three-way and oxidation catalysts were tested. The air/fuel ratio was controlled electronically and is characterized in the table by a "millivolt control point." Lower values represent leaner mixtures, and higher values richer mixtures. Stoichiometric conditions are represented by a value of about 550 mV.

To obtain adequate maximum power output, the carburetor control loop was opened at wide-open-throttle conditions to enrich the mixture. Such enrichment, however, raises oxidation catalyst temperatures and could limit catalyst life.

EGR was provided by the 1978 production system, which was deactivated for some tests and activated for others. All data in Table 21 reflect tests with EGR.

The data in Table 21 show that, with appropriate control of the air/fuel ratio (500-700 mV), the three-way catalytic converter can yield NO_x emissions of 1-2 g/bhp-h with fresh catalyst. However, for this particular installation, with air-fuel ratio adjustments in the indicated range, it was necessary to add a downstream oxidizing catalyst bed to control emissions of hydrocarbons and carbon monoxide.

Fuel economy for the three-way catalyst tests can be compared with base engine fuel economy data, also shown in Table 21. Operation at the richer settings, to minimize NO_x emissions, appears to increase fuel consumption by as much as 3-7 percent, though apparent test-to-test variability in fuel consumption measurements makes precise determination of fuel consumption trends difficult.

TABLE 20 Emissions of Heavy-Duty 350-Cubic-Inch-Displacement V-8 Gasoline Engine with Catalytic NO_x Control System

Engine Characteristics	Emissions and Fuel Consumption (g/bhp-h)			
	Hydrocarbons	Carbon Monoxide	NO_x	Fuel Consumption
1984 Prototype 350-CID V-8 with Dual Oxidizing Converters				
Without EGR	0.47	4.01	5.65	245
With EGR	0.67	5.97	2.91	254
1986 Prototype 350-CID V-8 with Three-Way Converters and EGR				
Full-time air	0.36	8.14	1.97	259
Modulated air	0.54	11.66	1.28	259

SOURCE: Informal communication, General Motors Corporation, February 4, 1981.

TABLE 21 Emissions and Fuel Consumption of Heavy-Duty Gasoline Engine with Three-Way Catalyst NO_x Control System (Fresh Catalyst)

Engine Configuration	Hydro-carbons	Carbon Monoxide	NO_x	Fuel Consumption	Increase in Fuel Consumption[a]
1978 Production	2.58	56.0	4.89	312	0
350 millivolts[b]					
Three-way catalyst with EGR	0.85	13.9	3.14	328	5.2
Three-way catalyst with oxidation catalyst and EGR	0.74	3.46	3.45	337	8.2
530 millivolts					
Three-way catalyst without EGR	1.41	22.9	1.58	320	2.6
Three-way catalyst with oxidation catalyst and EGR	0.64	3.21	1.48	321	2.9
630 millivolts					
Three-way catalyst with oxidation catalyst and EGR	0.56	3.29	0.71	335	7.3
740 millivolts					
Three-way catalyst with oxidation catalyst and EGR	0.68	3.60	0.74	319	2.3

Emissions and Fuel Consumption as Measured on Transient Cycle (g/bhp-h)

[a] Percentage change compared with 1978 production engine.

[b] Carburetor set point (350 mV = lean; 550 mV = chemically correct; 630 mV = rich).

SOURCE: Data from Hansel, Cox, and Nugent, 1981.

All of the test date of Table 21 were obtained with nearly fresh catalyst. Durability data for these systems are not available.

Catalyst Durability

As emphasized above, information on the durability of emission control systems is vital to assessing the feasibility of meeting given standards with catalytic converters. Unfortunately, the committee is not aware of any such data relevant to the proposed 1986 heavy-duty NO_x standards. Given this lack of information, it has been suggested that durability data on three-way catalyst systems in passenger cars could be applied to the heavy-duty situation. However, there is considerable data suggesting that catalysts in heavy-duty systems undergo much more severe operating conditions than those in light-duty systems.

A number of factors are responsible for the relative severity of heavy-duty service, but the most important by far is the potential for excessively high catalyst bed temperatures, leading to deterioration in catalyst activity. Such concern has been voiced with regard to deterioration of the oxidation catalysts required to meet 1984 hydrocarbon and carbon monoxide standards, as well as the three-way catalytic converters needed to meet the proposed 1986 NO_x standard. Since this report is concerned with the 1986 NO_x standard, this discussion will focus on three-way converters for NO_x control. However, the intent is not to minimize the potential severity of the durability problems of oxidation catalysts.

It is well known that exposure to excessive temperatures can permanently deactivate catalytic controls. For this reason, the higher exhaust system temperatures that may be encountered in heavy-duty vehicles are of major concern. Figure 13, taken from the previously cited EPA tests (Hansel, Cox, and Nugent, 1981), shows the maximum temperatures attained during the heavy-duty transient test procedure and during steady operation at maximum power (indicated by "MAP test"). Both tests subject the three-way converter to temperatures of 1400-1500°F. Temperatures encountered in actual vehicle operation may be even higher than those encountered in the engine dynamometer test, depending on operating conditions, ambient temperatures, and external air flow rates.

There appears to be no consensus of opinion on the maximum permissible three-way catalyst bed temperatures for acceptable durability. The referenced EPA paper states, "The maximum temperatures experienced by the [three-way catalyst] were satisfactory and within design limits." On the other hand, information submitted to this committee by the Ford Motor Company (informal communication, February 9, 1981) states that, "Temperature guide-lines for current [three-way catalysts] are 1300°F sustained and momentary spikes to 1450°F for rich operation. Future development efforts may permit somewhat higher temperature operation by 1986."

Catalyst-aging data obtained by General Motors are shown in Figures 14 and 15. In these tests, catalyst materials are aged in high-temperature furnaces with controlled atmospheres simulating the combustion products of a given stoichiometry. For fuel-lean conditions, catalyst durability is good for

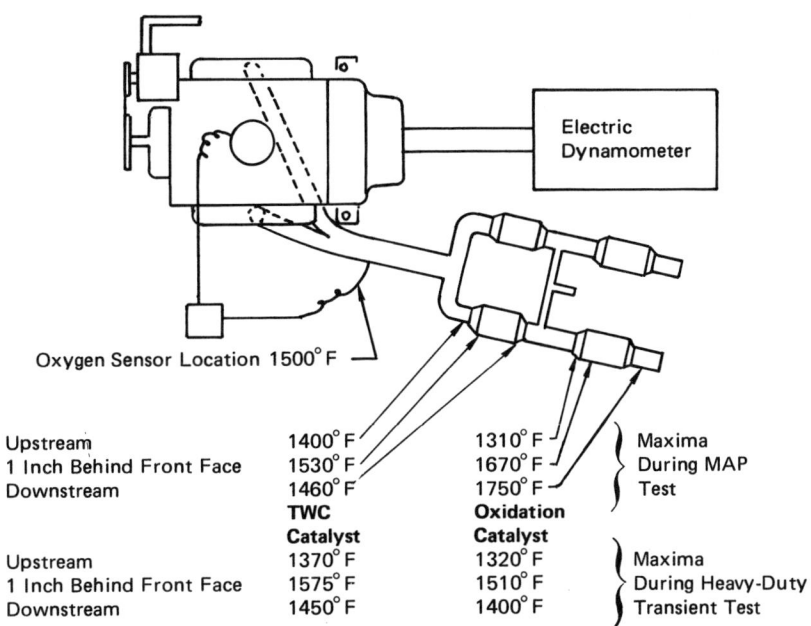

Figure 13 Maximum catalyst temperatures (Hansel, Cox, and Nugent, 1981)

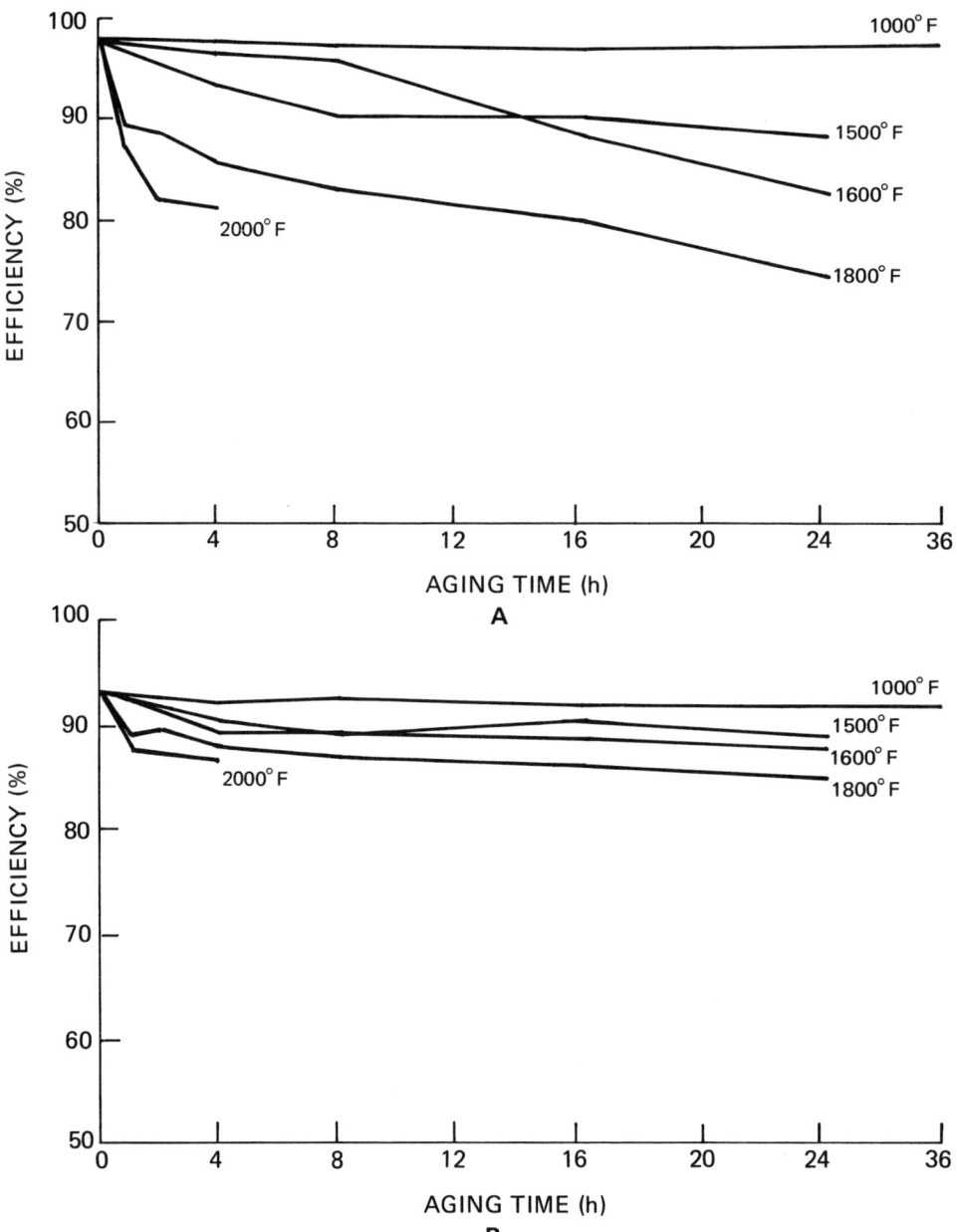

Figure 14 Effect of furnace aging on three-way catalyst hydrocarbon removal efficiency, after 600-second warmup: (A) in neutral atmosphere (stoichiometric + 0.1 air/fuel unit); (B) in strong oxidizing atmosphere (approximately 4-5 percent O_2). (Informal communication, General Motors Corporation, February 4, 1981.)

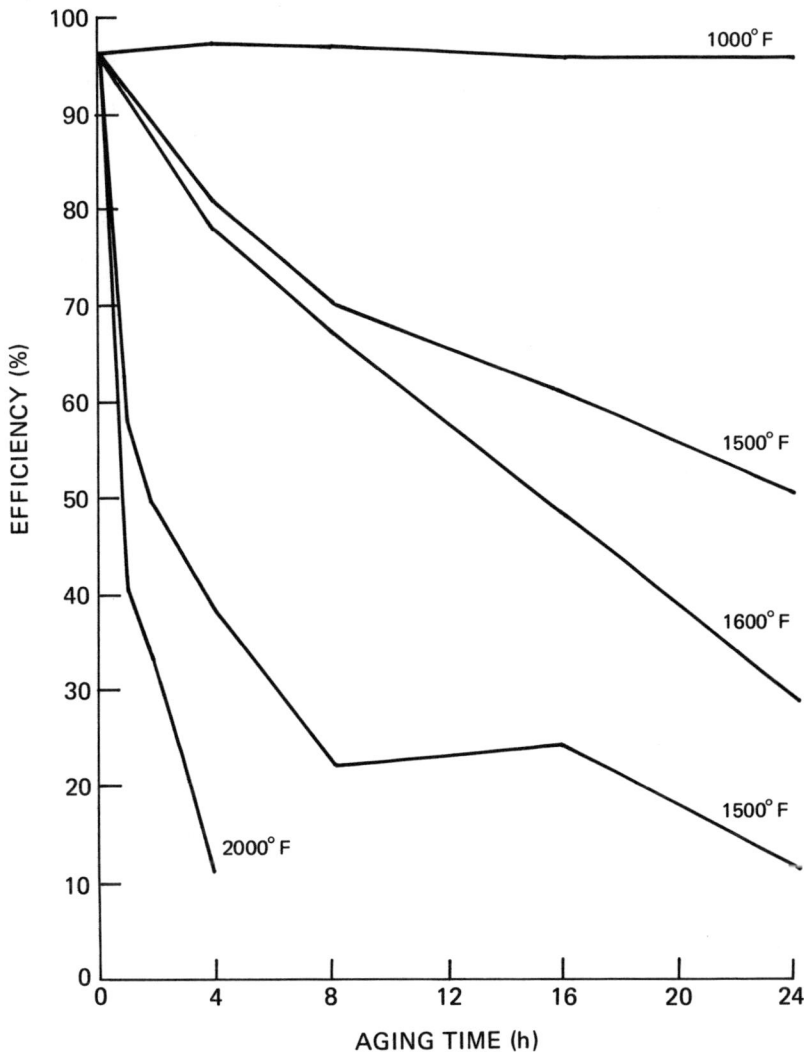

Figure 15   Effect of furnace aging of three-way catalyst NO_x removal efficiency, after 600-second warmup, in slightly rich atmosphere (stoichiometric + 0.2 air/fuel unit) (Informal communication, General Motors Corporation, February 4, 1981.)

temperatures as high as 2000°F. However, as the air/fuel ratio is enriched to chemically correct or slightly reducing conditions, high-temperature durability suffers substantially. For mixtures only 0.2 air/fuel units richer than chemically correct, catalyst activity is degraded significantly at temperatures as low as 1500°F.

Overall Assessment of Catalytic NO_x Control

As stated earlier, three-way catalytic converter systems can be expected to approach or meet the NO_x emission levels of the proposed 1986 emission standards with fresh catalyst materials. The critical question is catalyst durability in heavy-duty vehicle service. Of major concern is the catalyst's potentially severe operating temperature environment. Experimental data on the durability of three-way systems appear to be unavailable even at the laboratory engine level. While data on exhaust system temperatures are available, statements on maximum permissible bed temperatures are conflicting. Data supplied by one manufacturer, however, suggest that, for the reducing atmospheres employed in three-way catalyst systems, measured operating temperatures in heavy-duty service may exceed permissible levels.

To sum up, it appears that insufficient durability data are available currently to determine the feasibility of heavy-duty gasoline engines' meeting the proposed 1986 NO_x standard of 1.7 g/bhp-h by using three-way catalyst technology. Substantially more testing will be required, with a variety of emissions contol system configurations, before a meaningful assessment can be made.

The foregoing discussion has focused on the durability of NO_x reducing cataysts and has not considered a number of other development problems that must be addressed by engine manufacturers. Among these are the questions of satisfactory engine performance and durability (exhaust valve and valve seat life, octane satisfaction, maximum power output, fuel distribution, and driveability. In addition, manufacturers have expressed serious concerns about the durability of the oxidation catalysts used to meet 1984-1986 hydrocarbon and carbon monoxide emission standards. The intent has been not to minimize the importance of these concerns but rather to focus on the question that currently appears most critical to NO_x emission control in heavy-duty gasoline engines--the durability of three-way catalyst systems.

CONCLUSIONS

Evaluating the emission performance of new heavy-duty gasoline engines and control systems at low emission levels is made difficult by the very limited data available. For this reason, the emissions and fuel economy values cited in the following conclusions must be regarded as tentative. (All emissions data on the following are derived from the transient test procedure.)

NO_x emissions of about 5 g/bhp-h can be obtained in new gasoline-powered heavy-duty engines with little or no fuel economy penalty. The use of EGR combined with engine recalibration can yield new-engine NO_x levels of about 3 g/bhp, with fuel consumption penalties of about 3-7 percent. The higher EGR rates and retarded timing required for lower NO_x levels would further increase fuel consumption and at the same time reduce engine performance.

With fresh catalysts, one three-way catalytic control system has hydrocarbon, carbon monoxide, and NO_x emissions of 0.54, 11.66, and 1.28 g/bhp-h, respectively, with an accompanying 5-percent increase in fuel consumption relative to that of a prototype system with an NO_x level of 6.5 g/bhp-h. Also with fresh catalysts, one system with both three-way catalysts and oxidation catalysts has achieved corresponding emission levels of 0.6, 3.6, and 0.7 g/bhp-h, with increases in fuel consumption of 3-7 percent.

No data on the durability of catalytic NO_x control systems used with heavy-duty engines are available. However, data on exhaust temperatures indicate that catalyst durability may be a significant problem. For this reason, we cannot draw conclusions about the success of these systems until durability data become available.

REFERENCES

California Air Resources Board. 1981. "Public Hearing To Consider Amendments to Title 13, Section 1956.7, California Administrative Code, Regarding Exhaust Emission Standards and Test Procedures for 1984 and Subsequent Model Heavy-Duty Engines." Sacramento, Calif.: California Air Resources Board. (Staff report) January 21.

Hansel, James G., Timothy Cox, and Thomas Nugent. 1981. "The Application of a Three-Way Conversion Catalyst System to a Heavy-Duty Gasoline Engine." Warrendale, Pa.: Society of Automotive Engineers. (SAE Paper No. 810085.)

National Research Council. 1974. Report of the Committee on Motor Vehicle Emissions. Washington, D.C.: Society of Automotive Engineers. (SAE Paper No. 810085.)

Obert, E.F. 1968. Internal Combustion Engines. Scranton, Pa.: International Textbook.

Patterson, D. J., and N. A. Henein. 1972. Emissions From Combustion Systems and Their Control. Ann Arbor, Mich.: Ann Arbor Science.

Chapter 4
ENVIRONMENTAL HEALTH EFFECTS

This chapter provides some commentary on the health aspects of NO_x in relation to the regulation of exhaust emissions from heavy-duty vehicles. The intent is to place emissions control technology in the context of the health considerations. This chapter is not an original review of research in this field. Instead, the material is drawn largely from a number of published reviews of the health impacts of NO_x and diesel exhaust (National Research Council, 1977, 1981; U.S. Environmental Protection Agency, 1971; World Health Organization 1977). The discussion concentrates on the extent to which dose-response relationships can be defined at low levels of NO_x exposure and on the health trade-offs that may be involved in applying current control technologies (particularly the increased emissions of carcinogenic particulates with reductions in NO_x emissions).

BIOMEDICAL EFFECTS OF NO_x

The most toxic of nitrogen oxides is nitrogen dioxide (NO_2), and much of the concern with NO_x relates to the toxicity of NO_2. The effects depend on both concentration and duration of exposure (Gardner et al., 1979). The chemical reactions of NO_x leading to the toxic responses are essentially instantaneous and involve the formation of nitrous acid (HNO_2) and nitric acid (HNO_3). Both acids may be neutralized at rapid rates in the lung. NO_2 is only moderately water soluble and thus penetrates the the deep portions of the lung. The region of the terminal respiratory bronchials and adjacent alveoli are most affected (Coffin and Stokinger, 1977).

Damage from NO_2 exposure is almost completely limited to the lung, since the chemical reactions occur very rapidly. A major toxic effect is the death of respiratory tract cells. These cells are replaced, so that the net effect is a stimulation of cell turnover and some shift in cell types; for example, ciliated cells can be replaced by mucous-secreting cells. Biochemical indicators of injury in terms of, for example, increased cell permeability are observed with exposures as low as 0.4 ppm NO_2 (Menzel et al., 1977).

Pulmonary defenses against exogenous bacteria and viruses are diminished by even short-term exposures to NO_2. Single three-hour exposures at a concentration of 1 ppm NO_2 result in a small excess mortality in mice after challenge with infectious agents; recovery from this effect occurs in 24-36 hours (Coffin, Gardner, and Blommer, 1976).

Chronic exposure is also characterized by cell death and replacement of pulmonary cells. Biochemical and functional indicators of damage occur relatively rapidly and reach a steady state within about two weeks. The development of irreversible structural damage of the lung requires a long time and involves increased thickness of alveolar walls, loss of ciliated cells, narrowing of small airways, alterations in the morphology of clara cells, transformation of alveolar cells from Type II to Type I, and the appearance of altered alveolar collagen (Coffin and Stokinger, 1977). In rodents, the development of frank emphysema with extensive destruction of air spaces requires relatively high exposure (e.g., 15 ppm NO_2). The suppression of mortality from continuous exposures to high concentrations of NO_2 by anti-oxidants such as vitamin E and other free radical scavengers supports the possibility that the toxic effects of NO_2 are largely due to membrane damage by chemical oxidation of unsaturated fatty acids (Menzel, 1970). While there is some evidence for the development of tolerance from a biochemical standpoint, the possibility of adaptation to exposure to NO_2 is questionable at present. Additive toxicity with other air pollutants and NO_2 is most likely, but no evidence for potentiation has been found. The kinds of pulmonary damage caused by NO_2 and ozone are similar. In summary, the most sensitive toxicologic response to NO_2 in experimental animals appears to be a decrease in resistance to infection, and this has been observed at concentrations down to 0.5 ppm NO_2 for exposure times of 4 hours or more.

Direct evidence for the effects of NO_2 on humans comes from controlled exposure studies and from epidemiologic studies (National Research Council, 1977). Controlled exposure studies provide information on the effects of single short-term exposures. The most frequently observed effect of NO_2 exposure includes increases in airway resistance and changes in susceptibility to the effects of bronchial constricting drugs. There is reasonably clear-cut evidence for these effects at exposure concentrations above 2 ppm in healthy individuals with 5-15 minute exposures. A substantial series of studies fails to show any clear-cut evidence for effects on pulmonary function at exposures of 1 ppm or below. The functional significance of the induction of bronchial constriction or increased susceptibility to bronchial constricting drugs is not clear. There is some unconfirmed evidence for a greater than normal sensitivity to NO_2 in bronchitics, but the results of different studies are not consistent. The extent of hypersusceptibility in children, the elderly, and individuals with cardiovascular disease has not been studied.

Epidemiologic studies are complicated because there are usually complex mixtures of pollutants in the air. Epidemiologic studies of NO_2 effects made before 1973 are of questionable validity due to the use of the unreliable Jacobs-Hocheiser technique for measuring NO_2. With few exceptions, epidemiologic studies fail to show any significant effects on lung function in populations exposed to ambient levels of urban NO_2 pollution. There was evidence of an increased incidence of bronchitis in children in the Chattanooga, Tennessee, area who lived near a plant that emitted NO_2, but no reliable NO_2 estimates were associated with the several studies made of this population.

A number of epidemiologic studies have shown increased respiratory illness rates among young children living in homes using gas cooking stoves (Florey et al., 1979; Keller et al., 1979; Speizer et al., 1980). The annual average levels in these houses were on the order of 0.03-0.07 ppm, with peak levels during the operation of the stoves in the domain of 0.5-1.0 ppm. No such effect of indoor NO_2 pollution has been observed in adults.

HEALTH ASPECTS OF POLLUTANT TRADE-OFFS

Current methods of controlling NO_x emissions from heavy-duty diesel engines can increase emissions of carbon monoxide, hydrocarbons, and particulates. Heavy-duty diesels emit relatively small amounts of carbon monoxide, so that modest increases in emissions of this pollutant are not of special concern. Hydrocarbons interact with sunlight and NO_2 to form oxidants, which in general produce the same kinds of lung damage as NO_2. Hence, a rise in hydrocarbon emissions tends to vitiate the health benefits of reduced NO_x emissions.

The rise in particulate emissions with decreased NO_x emission is of special concern because diesel exhaust particles might constitute a carcinogenic risk to humans. Like the combustion of most organic materials, the combustion of diesel fuel produces polycyclic aromatic hydrocarbons (PAH), which are well-known carcinogens, as part of the output of particulates. These particulates have been shown to be mutagenic in a variety of assay systems and are capable of initiating skin cancer in mice (Pepelko, Danner, and Clarke, 1980).

Diesel exhaust has not been found, so far, to be carcinogenic when inhaled by laboratory animals; nor have epidemiological studies of various occupational groups revealed a convincing connection between diesel exhaust and human cancer (National Research Council, 1981). However, the available evidence, although negative, is consistent with a level of carcinogenic risk that is of serious concern. The scientific evidence clearly demonstrates the carcinogenicity of diesel exhaust constituents in laboratory animals. Further work, such as the Environmental Protection Agency's current research program to characterize the carcinogenic potential of diesel exhaust for humans, is needed.

For purposes of risk assessment, the dose-response relationships for carcinogens are generally assumed to be linear, and without thresholds of activity (Interagency Regulatory Liaison Group, 1979). The implication of this dose-response characteristic is that there is no such thing as a safe dose of a carcinogen; in other words, excess risks are induced by even minute exposures. This stems from the facts that mutagenicity and carcinogenicity both represent manifestations of genotoxicity and that there is ample evidence for the linear, nonthreshold character of mutagenesis. The initiation stage of carcinogenesis in mouse skin also shows a linear, nonthreshold pattern (Albert, 1981). Limited epidemiologic evidence also support the linear, nonthreshold character of carcinogenic action.

The dose-response relationships at low levels of exposure for NO_x are uncertain, but the rapid neutralization of nitric and nitrous acids by lung fluids would tend to make very low levels of NO_x exposure relatively nontoxic. Hence, there are grounds for believing, given the present state of knowledge, that the carcinogenic particulates in diesel exhaust may represent a more serious health risk than NO_x at low levels of exposure.

SUMMARY AND CONCLUSIONS

NO_x, and particularly NO_2, is toxic to the lung. The most important effect at concentrations below 1 ppm appears to be increased susceptibility to pulmonary infection. Recent epidemiologic evidence indicates that young children are susceptible at average levels not far from the current national air quality standard of 0.05 ppm, although this effect may in fact be due to transient spikes in indoor NO_2 concentrations, which can approach 0.5-1.0 ppm. There are no dose-response data for heightened susceptibility to pulmonary infection in animals at NO_2 concentrations below 0.5 ppm, and there is no dose-response data for this effect in humans. We have no knowlege of the mechanism of the effect in humans. For that reason, there is no basis for postulating the shape of the dose-response relationship. Consequently, considering NO_x in isolation, it appears to be a good idea to reduce atmospheric concentrations to the extent feasible.

However, there is no basis for an accurate prediction of the health impacts of increased NO_x levels on the order of 25-50 percent over 1976 urban levels, which have been projected to occur given a high-growth scenario for the use of heavy-duty vehicles without more stringent NO_x emission controls. In heavy-duty diesel engines, reducing NO_x emissions according to current control strategies may increase emissions of particulates. This represents a trade-off of very different kinds of health effects between the two classes of pollutants. The principal hazard from increasing diesel exhaust particulates is the possible excess risk of lung cancer; here, there is wide acceptance of a linear, nonthreshold dose-response relationship for the effects of low-level exposures to environmental carcinogens. By contrast, the effect of NO_x in heightening susceptibility to respiratory infection is probably a less serious toxic effect than lung cancer and, although the dose-response relationships for NO_x are highly uncertain, the likelihood that there is a threshold for NO_x is greater than that for the action of carcinogens. From a health standpoint, it might be imprudent to suppress NO_x emissions from heavy-duty engines at the expense of a substantial increase in the emission of particulates. The extent to which these two species affect human health requires further study, especially to assess the carcinogenic risk of diesel particulates.

REFERENCES

Albert, R. E. 1981. "A Biological Basis for the Linear Non-Threshold Dose-Response Relationship for Low-Level Carcinogen Exposure." In Measurement of Risks (G. Berg and H. David Maillie, eds.). New York: Plenum Publishing Corp.

Coffin, D. L., and H. E. Stokinger. 1978. "Biological Effects of Air Pollutants." In Air Pollution, 3d ed., vol. 2 (A. C. Stern, ed.), p. 264-351. New York: Academic Press, Inc.

Coffin, D. L., D. E. Gardner, and E. J. Blommer. 1976. "Time-Dose Response for Nitrogen Dioxide Exposure in an Infectivity Model System." Environmental Health Perspectives 13:11-15.

Florey, C. du V., R. J. W. Melia, S. Chinn, B. D. Goldstein, A. G. F. Brooks, H. H. John, I. B. Craighead, and X. Webster. 1979. "The Relation Between Respiratory Illness in Primary School Children

Gardner, D. E., F. J. Miller, E. J. Blommer, and D. L. Coffin. 1979. "Influence of Exposure Mode on the Toxicity of NO_2." Environmental Health Perspectives 30:23-29.

Interagency Regulatory Liaison Group, Work Group on Risk Assessment. 1979. "Scientific Bases for Identification of Potential Carcinogens and Estimation of Risks." Journal of the National Cancer Institute 63:241-268.

Keller, M. D., R. R. Lanese, R. I. Mitchell, and R. W. Cote. 1979. "Respiratory Illness in Households Using Gas and Electricity for Cooking." Environmental Research 19:495-503.

Menzel, D. B. 1970. "Toxicity of Ozone, Oxygen, and Radiation." Annual Review of Pharmacology 10:379-394.

Menzel, D. B., M. D. Abou-Donia, C. R. Roe, R. Ehrlich, D. E. Gardner, and D. L. Coffin. 1977. "Biochemical Indices of Nitrogen Dioxide Intoxication of Guinea Pigs Following Low-Level Long-Term Exposure." In Proceedings of the International Conference on Photochemical Oxidant Pollution and Its Control, September 1973 (B. Dimitriades, ed.). Research Triangle Park, N.C.: U.S. Environmental Protection Agency. (EPA 600/3/-77-001b)

_____. 1981. *Health Effects of Exposure to Diesel Exhaust*. Diesel Impacts Study Committee, Health Effects Panel. Washington, D.C.: National Academy Press.

Pepelko, W. E., R. M. Danner, and N. A. Clarke, eds. 1980. *Health Effects of Diesel Engine Emissions: Proceedings of an International Symposium*, vol. 2. Cincinnati, Ohio: Health Effects Research Laboratory. (EPA 600/9-80-057b)

Speizer, F. E., B. Ferris, Jr., Y. M. M. Bishop, and J. Spengler. 1980. "Respiratory Disease Rates and Pulmonary Function in Children Associated with NO_x Exposure." *American Review of Respiratory Diseases* 121:3-10.

U.S. Environmental Protection Agency. 1971. *Air Quality Criteria for Nitrogen Oxides*. Air Pollution Control Office. Washington, D.C.: U.S. Environmental Protection Agency.

World Health Organization and U.N. Environment Program. 1977. *Oxides of Nitrogen*. Geneva: World Health Organization.

Chapter 5
CONTROL COSTS AND OTHER REGULATORY QUESTIONS

This chapter presents some regulatory issues that arise from the information on technology and health effects developed in previous chapters. The costs associated with this regulation will be discussed first. The conclusions of this chapter will focus on questions that need to be answered as part of any legislative or regulatory consideration of NO_x emission controls on heavy-duty engines.

CONTROL COSTS

In a questionnaire, the committee asked engine manufacturers to estimate the costs imposed by tighter NO_x standards in three categories: (1) hardware and initial testing, (2) any increased maintenance, and (3) the increased fuel consumption associated with emission control.

The estimates of capital costs were very tentative, especially for heavy-duty gasoline engines. No information was received on maintenance costs, but these are expected to be small compared to the other cost categories. It became obvious, however, that any significant increase in fuel consumption would have costs greatly outweighing the initial capital costs. Thus, the consideration of the costs of NO_x regulations must focus on the change in fuel consumption. Table 22 shows the information developed by EPA in its regulatory analysis.

These numbers were based on particular assumptions about fuel economy and annual mileage, but other reasonable assumptions would show the same general relation between initial cost and fuel penalty. For diesel engines the fuel cost of a 1-percent increase in fuel consumption is roughly equal to the initial capital cost increase. For gasoline engines, a 2-percent increase in fuel consumption gives a fuel cost increase approximately equal to the initial capital cost. In the remainder of this section we will consider only fuel costs. It is important to emphasize that the costs given in this section are illustrative only. Although their approximate magnitudes are known, and the conclusions we draw from them sound, actual fleet average data on fuel penalties and capital costs are necessary for accurate cost-effectiveness calculations. We have not evaluated, even in a qualitative fashion, the total benefits of NO_x control. Nor have we provided any kind of cost-benefit analysis. These should be considered parts of a full regulatory analysis.

TABLE 22 EPA Cost Estimates for NO_x Controls

Cost Category	Cost (Dollars)	
	Gasoline Engines[a]	Diesel Engines[b]
Capital cost[c]	270	733
Cost of 1-percent increase[d] in fuel consumption	150	754

[a]Control technology: three-way catalysts substituted for oxidation catalysts.

[b]Control technology: turbocharging, charge cooling, electronic injection timing controls, and exhaust gas recirculation.

[c]Undiscounted costs for hardware, R&D, and certification testing.

[d]Lifetime fuel costs, undiscounted.

SOURCE: Adapted from U.S. Environmental Protection Agency (1980).

We have evaluated some typical data on heavy-duty truck use to estimate the fuel consumption costs for different vehicle weight classes (Table 23). Specifically, the costs per mile of a 1-percent increase in fuel consumption ranges from 0.12 cents (for diesels in weight classes 3-5) to 0.40 cents (for gasoline engines in classes 7-8). The same 1-percent increase in fuel consumption increases the lifetime fuel cost by $258 to $1,151. Of course, the costs of a 1-percent fuel penalty can simply be multiplied by the actual percentage increase in fuel consumption to obtain the actual costs.

The implications of increased fuel consumption on the entire heavy-duty truck fleet are illustrated in Table 24. The projected fuel use is for 1995, when most of the fleet miles will be driven by trucks produced in the mid-1980s and later. The data are from the base case in a paper by Jambekar and Johnson (1981). The fuel use for class 2B (8,501-10,000 pounds gross vehicle weight rating) was estimated as 5.5 percent of the total light truck fuel use. The figure of 5.5 percent was taken from data in the EPA regulatory analysis (U.S. Environmental Protection Agency, 1980), as the percentage of light trucks in class B.

Table 24 shows that a 1-percent increase in fuel consumption for the heavy-duty fleet leads to a total annual cost of $430 million per year; $397 million of this is for diesel fuel, of which $338 million is for class 7 and 8 trucks.

From a regulatory perspective, one important number is the cost-effectiveness of a given degree of control (that is, the dollar cost divided by the emissions reduction). A proper cost-effectiveness accounting must consider the sales-weighted increase in fuel consumption for all vehicles in the fleet. This requires a detailed analysis that is beyond the scope of the committee's charge. To illustrate the relative cost-effectiveness of various controlled emissions levels for diesel and gasoline engines we have calculated the increase in fuel costs for assumed fuel penalties and assumed emissions levels. These cost-effectiveness numbers (for fuel costs only) are shown in Table 25. Although the percentage fuel penalty, at a given NO_x level, is likely to be different for the two different engine types, Table 25 shows that the cost-effectiveness is better for the diesel when the penalty is the same for both engines. The cost-effectiveness for both engines is the same only when the percentage fuel penalty of the diesel is greater than that of the gasoline engine. There are two reason for this. First, control to a given level implies greater emissions reductions for diesel engines than for gasoline engines. Second, for a given percentage increase in fuel consumption, the more efficient diesel engine increases its actual fuel consumption less than the gasoline engine.

INDIRECT COSTS OF NO_x CONTROL

In assessing the cost of an NO_x control regulation, it is important to consider increases in any other pollutant species cause by the control of NO_x. The diesel engine chapter describes the increases in particulate and hydrocarbon emissions associated with increases in the stringency of NO_x emission control.

TABLE 23 Fuel Costs for Typical Heavy-Duty Vehicles

	Data by Weight Class					
	Classes 3-5		Class 6		Classes 7&8	
	G	D	G	D	G	D
Gasoline or diesel	G	D	G	D	G	D
Assumed lifetime (years)	8	9	8	9	8	9
Lifetime mileage[a] (thousands of miles)	127	247	126	247	137	503
Fuel economy in 1986[b]	6.84	10.88	6.03	9.77	3.49	5.90
Lifetime fuel use (thousands of gallons)	18.4	22.7	20.9	25.3	39.3	85.3
Assumed fuel cost (dollars per gallon)	1.40	1.35	1.40	1.35	1.40	1.35
Fuel cost/distance traveled (cents per mile)	20.5	12.4	23.2	13.8	40.1	22.9
Lifetime fuel cost (thousands of dollars)	25.8	30.6	29.3	34.2	55.0	115.1
Cost of a 1% Increase in Fuel Consumption						
(cents per mile)	0.205	0.124	0.232	0.138	0.401	0.229
(lifetime cost, in dollars)	258	306	293	342	550	1151

[a]Data from Energy and Environmental Analysis, Inc. (1980).

[b]Data from Jambekar and Johnson (1981), Table 9.

TABLE 24 Cost Impact of Fuel Penalties on Heavy-Duty Fleet in 1995 (Annual Costs)

Class	Gasoline Engines, by Class				Gasoline Total	Diesel Engines, by Class				Diesel Total
	2B[a]	3-5	6	7 and 8		2B[a]	3-5	6	7 and 8	
Fuel (billion gallons)[b]	1.1	0.3	0.9	0.1	2.4	0.4	0.2	3.8	25	29.4
Fuel Cost (billion dollars)[c]	1.5	0.4	1.3	0.1	3.3	0.5	0.3	5.1	33.8	39.7
Cost of a 1% fuel penalty (millions of dollars)	15	4	13	1	33	5	3	51	338	397

[a]Assumed as 5.5 percent of total light-duty truck fuel

[b]Data from Jambekar and Johnson (1981)

[c]Assumed fuel costs: gasoline $1.40/gallon; diesel $1.35/gallon

TABLE 25 Fuel Costs at Selected Assumed NO_x Emission Rates and Fuel Consumption Penalties[a]

Engine Type	Fuel Consumption Penalty, (percent)	Cost (dollars) per Ton of NO_x Abated at NO_x Emission Rates (g/bhp-h) of:					
		7	6	5	4	3	2
Gasoline							
	1			1,400	700	470	350
	2			2,800	1,400	930	700
	5			7,000	3,500	2,300	1,750
	10			14,000	7,000	4,700	3,500
	20			28,000	14,000	9,300	7,000
Diesel							
	1	900	450	300	225	180	150
	2	1,800	900	600	450	360	300
	5	4,500	2,250	1,500	1,125	900	750
	10	9,000	4,500	3,000	2,250	1,800	1,500
	20	18,000	9,000	6,000	4,500	3,600	3,000

[a]The assumed data are as follows:

- Current brake-specific fuel consumption (BSFC) 200 g/bhp-h for diesel engines, 300 g/bhp-h for gasoline engines

- Current NO_x emissions (E_1) 8 g/bhp-h for diesel engines, 6 g/bhp-h for gasoline engines

- Fuel costs (C_f) $1.35 for diesel fuel, $1.40 for gasoline

- Fuel density (P_f) 6 lb per gallon.

The cost-effectiveness (CE) for a given controlled emission level E_2 and percentage fuel penalty P is given by the equation

$$CE = \frac{(P/100)(BSFC)(C_f P_f)}{E_1 - E_2}$$

With the above data and the necessary unit conversion factors, this equation yields the following computational equations:

$$CE = \frac{1400\ P}{6 - E_2}$$

for gasoline engines, and

$$CE = \frac{900\ P}{8 - E_2}$$

for diesel engines. Units are dollars per ton of NO_x emissions abated.

Emissions of particulates, hydrocarbons, and carbon monoxide are governed by existing or proposed regulations; any NO_x control technique that would increase these emissions above the regulated level would therefore necessitate additional control technology, and thus additional costs.

Emissions of unregulated pollutants (e.g., odor or sulfates), which might also increase as a result of more stringent NO_x control, could also be considered as imposing a cost in the form of additional health or welfare effects on the public.

The committee has no quantitative information on the costs that can be attributed to emissions increases resulting from NO_x control techniques. It is often difficult to determine how much of the emission control cost can be attributed directly to a specific pollutant species when the engine has been designed as a total package, which must meet all applicable standards.

The capital costs of NO_x control are difficult to assess. Particular engine design changes, for example, may improve performance and reduce fuel consumption in addition to controlling emissions; in such cases there is no simple way to say how much of the costs should be attributed to emission control. (Certification testing adds an additional capital cost, which is small compared with other cost elements.) Despite this uncertainty, which gives a range of $700 to $2,000 as estimates for the capital costs, per engine, of NO_x controls, the easily calculated costs of any increase in fuel consumption due to NO_x control become the largest component of the control costs if the corresponding increase in fuel consumption is greater than about 2 percent.

REGULATORY ISSUES

This section is not intended to be a full regulatory analysis. The questions discussed here arose in the course of the committee's study, but answering them will require work by the appropriate regulatory or legislative bodies. In addition, other important issues (e.g., the financial health of companies affected by the regulations) are not discussed here, but should be considered in developing regulations.

Differences Between Gasoline and Diesel Engines

The differences between gasoline and diesel engines have been described in previous chapters. From an emission control standpoint, the diesel engine has lower emissions of hydrocarbons and carbon monoxide, while the gasoline engine has lower NO_x and particulate emissions. Catalytic emission control systems are widely used on light-duty gasoline engines, but their durability under the more severe operating conditions of heavy-duty engines remains to be proven. A catalytic system cannot be used on a diesel engine unless a supply of a reducing agent such as ammonia or methane is carried on board the vehicle to react with NO_x. As noted in the introduction to this chapter, homogeneous reactions for NO_x removal are not possible at the temperatures and compositions of diesel exhaust. Diesel engines will find it harder than gasoline engines to achieve low NO_x emissions without a large fuel penalty.

On the other hand, because diesel engines are the largest fuel users of the heavy-duty engine fleet, any emission standard that imposes a fuel consumption penalty will have a greater effect on diesel users than on gasoline engine users.

From a regulatory perspective, the cost of the increased fuel use required to reduce a unit amount of NO_x emissions from diesels appears to be less than the comparable cost for gasoline engines. Given the differences between gasoline and diesel engines, it is appropriate to consider separate emission standards for the two engine types. In considering this question, it is necessary to analyze the relative competitive advantages of the two engines. There is already a trend toward the use of diesel engines in the lower weight ranges of the heavy-duty vehicle classes. This trend could be accelerated if separate standards improved the competitive advantage of the diesel (for example, by imposing heavier fuel consumption penalties on gasoline engines).

The net impact of separate standards for diesel and gasoline engines, taking into account effects on emissions, the issue of relative competitive advantage, user costs, and the cost-effectiveness of emission controls, will be difficult to analyze. But such separate standards should be considered, if only because of the simple fact that it is easier to control the NO_x emissions of gasoline engines than those of diesel engines.

Vehicle Size Considerations

The use of separate standards for the heaviest heavy-duty trucks (those in classes 7 and 8) has been suggested. As noted in the introduction, only 17 percent of the fuel used by these vehicles is used in urban areas, where emissions are most harmful. In addition, vehicles in these size ranges account for nearly three-fourths of the fuel used in heavy-duty vehicles. Thus, it appears that a higher NO_x emission standard for these vehicles would have a small impact on urban air quality and would increase the total fuel consumption of the heavy-duty fleet less than a uniform regulation. Any analysis of separate standards for different vehicle sizes must consider the impact of their emissions on total air quality. The introductory section on emissions inventories reported a forecast of the California Air Resources Board that heavy-duty diesels (most of which are in classes 7 and 8) would contribute 24 percent of the total NO_x emissions in the South Coast Air Basin in 1987. This is only one area of the country, but it is the one with the highest annual average NO_x concentrations.

Any consideration of less stringent NO_x standards for class 7 and 8 vehicles must examine the emissions that will not be reduced because of these less stringent standards, the impact of these foregone reductions on air quality, and alternative control strategies for obtaining the desired emission reductions.

Emissions Averaging

"Emissions averaging" is a regulatory concept under which the emissions of a given population of engines are averaged for purposes of regulation, so that some may exceed the prescribed average emission rate so long as enough others have emissions lower than the average. There are several emissions averaging schemes, differing according to the populations over which the averages are taken. For example, the current system, under which each engine family from each manufacturer must meet the emissions standards on an average basis, determined by the regulations for acceptable quality limit (AQL), might be called emissions averaging, since a manufacturer can produce individual engines that do not meet the standards. However, what is generally meant by the term is averaging over a larger population, such as the entire output of an engine manufacturer, or even an industry.

The concept of corporate averaging, which is currently used in the fuel economy regulations for passenger cars, has been suggested for emissions standards. The fuel economy regulations require that each manufacturer meet a corporate average fuel economy (CAFE) standard for its entire production. Under the corporate average emissions standard concept, a manufacturer could produce an engine family whose entire production exceeded the emissions standard provided that the same manufacturer also produced another engine family with emissions correspondingly below the standard.

In concept, emissions averaging in heavy-duty engines could be applied as follows: Production of diesel engines would be used to meet the hydrocarbon and carbon monoxide standards and production of gasoline engines to meet the NO_x and particulate standards, in recognition of the differences between the two engine types in terms of the difficulty of controlling emissions of each type. Production of the two engine types would be balanced to meet corporate average emission standards. If this were not possible, suitable controls could be placed on the engines whose emissions could be limited in the most cost-effective manner.

The difficulty with this concept is that only two manufacturers (General Motors and International Harvester) produce both types of engines. EPA has suggested that manufacturers of different engine types could work cooperatively in reducing the emissions from their two (or more) companies to meet the average emissions standards. If this could be done, a form of the corporate averaging concept could be used. However, its use would be limited by the relative demands for the two types of engines. An alternative to averaging the emissions of the two engine types is the use of separate standards for gasoline and diesel engines, as discussed earlier.

Another form of averaging is one that applies only to a single engine type. Here corporate average standards could be set for diesels and for gasoline engines; these standards could be the same or they could be different. Manufacturers could then determine which of their product lines could be controlled in the most cost-effective fashions. Since diesels would be averaged with diesels and gasoline engines with gasoline engines, this form of averaging would not affect the relative competitive advantages of the two engine types.

The U.S. Environmental Protection Agency (1981a) has announced that it will propose regulations which provide for averaging when it issues the final regulations for NO_x emissions of heavy-duty vehicles, in May 1982. The agency has not yet indicated which type of averaging it will propose. A fleet-average emissions standard could allow manufacturers to use emission-control technologies on only parts of their engine lines. This would provide in-use experience with new technology without the increased risk incurred when new technology is introduced over an entire product line. A fleet-average emissions standard that accomplished this could have a significant positive effect on the development of technology.

Test Procedures

Emissions reductions achieved in practice depend on the laboratory test procedures used to measure engine emissions. These test procedures should be reviewed continuously to ensure that emissions reductions measured in the laboratory are representative of actual, on-the-road emissions reductions from heavy-duty vehicles.

The transient test cycle that EPA has mandated for heavy-duty emissions standards in 1984 and subsequent model years represents a major change from the existing heavy-duty test cycle. (Appendix C describes and compares the old and new cycles.)

In adopting the cycle (U.S. Environmental Protection Agency, 1980a) EPA noted criticisms regarding the "justification for the tests, their representation of real life operation, their validation, their repeatability, and the lack of current knowledge upon which to base comments. After evaluating the comments EPA concluded that the test procedures were "necessary and appropriate." It further stated that:

> The origin of the operating cycles was an extensive program of actual in-use operational data collection and we are confident that the emphasis placed on quality during the subsequent cycle development program assures real world operating characteristics are well represented in the laboratory. It will not be possible to validate the cycle through time-consuming comparisons between on-the-road emissions and laboratory emissions. But, the pains taken to assure the representativeness of the cycles, in our view, would make such a validation superfluous anyway.

EPA also concluded that all labs running the transient tests at the time the regulations were promulgated showed a "degree of correlation."

The committee has received many criticisms of the cycle from engine manufacturers who disagree sharply with the conclusions of EPA. These disagreements are the subject of a current lawsuit by engine manufacturers against EPA. This committee has not conducted a detailed analysis of the manufacturers' objections to the test cycle or of the EPA response to these objections; we offer no conclusions regarding the validity of arguments on either side of this issue. Our comments are directed to further considerations of the test procedures. EPA has been asked to determine whether the 1984 heavy-duty truck requirements should be further revised based on the results of the manufacturers' current heavy-duty transient test programs (Executive Office of the President, 1981).

Any test procedure for mobile source emissions represents a compromise between the need for an accurate representation of on-the-road emissions (which might dictate a long and complicated test procedure) and the need for a test procedure that can be easily and inexpensively performed by manufacturers and control agencies. Any further evaluation of the test procedures by EPA should consider the appropriateness of test procedures with regard to this necessary compromise.

It is also important to consider the accuracy and repeatability of the test cycle. In a technical paper (Cox, 1980), EPA has noted that "additional throttle calibration specifications for throttle controller performance are necessary to assure repeatability" for the gasoline engine test procedure. The same paper concluded that "an additional tightening of the calibration 'window' for allowed torque variations appears necessary to assure repeatibility." This indicates EPA's own recognition of the need for review of the test procedure. Table 12, in Chapter 2, noted the variation in emissions for the same engine when tested at different labs. It is important to have more cross-checks of this type to ensure the accuracy of the data used in any regulatory analysis.

The ability of the test procedures to represent actual emissions performance should be subjected to an ongoing review.

Regulations and Technological Feasibility

Some recent environmental regulations have set standards, to take effect some time after promulgation, that could not be met by the existing technology. Such regulations (among them the proposed NO_x emission regulations for heavy-duty vehicles), are called "technology forcing," and are often very controversial. Their use by legislative and regulatory bodies is based on the primacy of the health goal the standard seeks to achieve and the assumption that the regulation can prompt the development of the desired technology. There is often uncertainty about whether such standards can actually be met, and this uncertainty is of great concern to companies affected by the standard.

The success of such regulations depends on a number of factors. The regulation should be perceived as firm by the industry being regulated. Regulations that are not firmly grounded technically, procedurally, or analytically may be perceived as likely to change; this perception impedes compliance efforts. Long-term consistency of approach on the part of the regulators is especially important to firms with limited technical or financial resources, who may not be able to undertake costly development programs.

It is also important that the regulatory requirements be consistent with the general momentum in the industry. For example, fuel economy standards have been achieved and surpassed because of the market forces that reinforced them. Conversely, regulations that contradict market trends or are so onerous as to force companies out of business are strongly resisted.

Another factor in technology-forcing regulations is the possibility that additional companies may enter the emissions control business in response to the standard. Technology-forcing regulations must provide long lead times if they are to achieve significant technological change. The establishment of such regulations is a clear signal to other companies that a market is present if they can develop the necessary technology.

In the specific subject considered here, the control of NO_x emissions from heavy-duty engines, two clear problems would be raised by the establishment of a technology-forcing regulation: the durability of catalysts for control of NO_x from gasoline engines and the effects of exhaust gas recirculation on diesel engines' performance and durability. In each of these cases, technology development by industry would be influenced by the level of the standard that is ultimately set.

This interaction between the setting of the standard and the technology that might become available to meet it has not been discussed by the committee. We have identified technologies likely to be available, based on an analysis of the current state of technology and of technology now under development. The question of what sort of technological developments will be promoted by what levels of standards is one that must be determined in the regulatory process.

TECHNOLOGY BEYOND 1986

This study has focused on heavy-duty emission control technology to be available in time to meet the Clean Air Act's 1986 deadline for promulgation of an NO_x standard for heavy-duty vehicles. Chapters 2 and 3 note the pervasive uncertainty about future technology, an uncertainty taken account of in the Act's provisions for the setting of interim standards to be revised every three years.

Because it is possible for the Administrator of the Environmental Protection Agency to set an interim NO_x standard less stringent than the 75-percent reduction specified in the Act, and because the Act itself is currently undergoing review, it is important to consider technology that could become available over a longer period of time. The technologies discussed in this report will continue to evolve. Some technologies whose availability by 1986 is uncertain will be available, for example, by 1990. New engine designs, using improved technology, will still exhibit certain trade-offs, but they can provide improvements in all aspects of engine operation--performance, emissions, and fuel economy--compared to current engines.

Improved Fuel Controls

Electronic fuel controls and improved fuel handling and injection systems for diesel engines are perhaps the most significant new technology likely to be widely available by the end of this decade. These systems are being developed now, and should be available on some production engines by the mid-1980s, yielding operational experience that will point the way toward improvements in design.

Introduction of these systems over a period of years should allow manufacturers the time they need to alter their engine designs to take advantage of the new systems and their electronic controls. As discussed in Chapter 2, the main effect that improved fuel controls will have on NO_x emissions will come from allowing the use of other control techniques (e.g. EGR) while minimizing the deleterious side effects encountered with current injection systems.

Catalyst Technology

The durability of catalysts for NO_x emission control in heavy-duty gasoline engines has also been noted as a source of uncertainty. These catalysts have been shown to be effective when fresh. Additional tests of their durability will determine whether they can remain effective in actual heavy-duty engine operation. If not, it will be necessary to develop new catalyst formulations capable of sustaining the high exhaust temperatures encountered in heavy-duty gasoline engines. This testing and development should be completed in time for catalysts to be used on heavy-duty gasoline engines by 1990.

The use of catalysts for removing NO_x from diesel engine exhaust is possible if a suitable reducing agent, such as ammonia, is added to the exhaust stream. One manufacturer who has studied such a system notes that a 50-percent reduction in engine-out NO_x emissions can be thus obtained. Problems of catalyst durability, plugging, and catalytic performance over a sufficiently wide temperature range have been noted. This kind of system is a longer term option, which may be available in the 1990s. Ultimately, the use of such a system depends not only on its technological development, but also on the regulatory judgment that vehicle operators would faithfully refill the tank containing the reducing agent, when this action would have no perceptible effect on engine performance.

Exhaust Gas Recirculation

The extent to which exhaust gas recirculation (EGR) will be used in heavy-duty diesel engines by 1986 is also uncertain. This technology too is undergoing development and should be generally available by 1990. One of the major problems with EGR is its likely effect on engine durability. The use of electronic controls to program EGR systems should improve this situation by allowing EGR rates to be varied so that the net effect is the greatest possible reduction in NO_x emissions at the minimum effect on durability. Such programmed EGR systems should be available for widespread use by 1990.

Particulate Traps

It has been observed in this report that with current control technology a decrease in NO_x emissions from diesel engines generally implies an increase in emissions of particulates. However, it would be possible to reduce emissions of both types simultaneously if exhaust particulate traps were available. A number of firms are attempting to develop such traps, mainly for light-duty diesel engines (National Research Council, forthcoming). The Environmental Protection Agency (1981a) believes that these systems will be available for heavy-duty diesel engines in time for compliance with a 1986 standard; this committee, however, is uncertain of this. Still, because of the variety of development programs in industry, it is quite possible that this technology will be available for use in 1990 vehicles. (Although this technology would not directly reduce NO_x emissions, it would allow the use of NO_x control techniques that might otherwise be ruled out by their effects on particulate emissions.)

Alternative Fuels and Engines

New types of engines and fuels may be significant in the control of NO_x emissions by the 1990s, but it is clear that conventional engines and fuels will be used in almost all heavy-duty applications for at least the next decade.

The use of alternative fuels in diesel engines is the subject of a recent Society of Automotive Engineers (1981) compendium. Some commonly mentioned alternatives are alcohols and various emulsions (such as water or alcohols with diesel fuel). The use of such fuels will require not only the development of new engines that can burn them properly, but also the establishment of fuel manufacturing and marketing infrastructures. Despite these obstacles, the use of alternative fuels in diesel engines is receiving a great deal of attention because of the possibility that they can be burned with low particulate emissions (U.S. Environmental Protection Agency, 1981c). The status of these fuels for use in light-duty engines is reviewed in a National Research Council (forthcoming) report. The possibilities of lower emissions, as well as the attractions of potential alternatives to petroleum, should provide an impetus for further research in this area.

The use of alternative engines is another possible means of reducing emissions over the long term. The forthcoming National Research Council report discusses alternative engine types in some detail. Although that report evaluates these engines for their applications in light-duty vehicles, its comments on the various engine types are generally valid for heavy-duty engines as well.

Two developments are of particular interest for heavy-duty vehicles. United Parcel Service (UPS) has contracted with Texaco and Ricardo Consulting Engineers to develop a stratified-charge version of their current General Motors gasoline engines, used in delivery vans.* Texaco has had this engine under development since the early 1940s. If successful, the engine could replace the engines in about three-fourths of the current UPS fleet. UPS has chosen this approach because its experience with diesel engines in delivery vans has been unsatisfactory despite fuel savings as high as 20 percent. Previous versions of the Texaco stratified charge engine, in light-duty vehicles, have shown good fuel economy with moderate emissions, but stringent emission controls sharply increased its fuel consumption (Tierney et al., 1975). The UPS sees its planned engine as a possible way to cut fuel consumption without the high initial cost of a diesel engine. Further work will show what kinds of emissions the engine can achieve.

Another promising alternative engine for heavy-duty applications is the gas turbine being developed by a consortium of the Garrett Turbine Engine Company, the Mack Truck Company, and KHD (a German company).** This is a very large engine, rated at 550 horsepower in commercial applications. Its emissions of hydrocarbons, carbon monoxide, and nitrogen NO_x have been measured at 0.05, 1.89, and 3.13 g/bhp-h, respectively, on the steady-state test cycle. At its most efficient operating point it has a brake-specific fuel consumption of 0.393 lb/bhp-h. Its particulate emissions have been reported as 0.33 grams per kilogram of fuel. Using the best fuel economy figure of 0.393 lb/bhp-h gives a minimum brake-specific particulate emission rate of 0.38 g/bhp-h. (Of course, the actual emission rate over the test cycle would be greater than this, and no direct comparison with the proposed 1986 particulate standard of 0.25 g/bhp-h, on the transient test procedure, is possible.) The developers of this engine are optimistic about its introduction in the latter half of the 1980s, but they recognize that this high-powered engine's uses will be limited. Initial applications are expected to be in on-and-off-road applications such as logging and mining. The first application in trucks would be in class 8 trucks operating in rugged terrain. This particular engine is cited here because it is currently undergoing on-road evaluations in a truck. Future use of this engine will obviously depend on its ability to meet NO_x and particulate standards, as well as on its fuel economy as compared to that of the diesel engines it would replace.

Gas turbines and stratified charge engines are the most well developed of all alternative engines. Their ultimate use will depend on further development to reduce fuel consumption and emissions. The levels at which emissions standards are set in the future will obviously play a part in the development of these or any other alternative engine technologies.

---

*Ruth A. Hunter, Massachusetts Department of Transportation, letter to Dennis Miller, January 20, 1981.
**T. W. Qualle, Garrett Turbine Company, letter to Laurence S. Caretto, April 21, 1981.

CONCLUSIONS

Although data on the costs of NO_x controls are preliminary and tentative, it is clear that any increase in fuel use of greater than about 2 percent, due to the imposition of NO_x emission controls, will provide the greatest component of the control cost. This issue needs the most attention in determining the costs of compliance with an NO_x control regulation. Issues that should be considered during the regulatory processes include:

o The differences between gasoline and diesel engines

o The different size ranges in heavy-duty engines

o Emissions averaging concepts

o The amount of research and development that would be promoted by a specific standard

o The ability of the industry to respond to the regulation.

REFERENCES

Cox, T. P. 1980. "Heavy-Duty Gasoline Engine Emission Sensitivity to Variations in the 1984 Federal Test Cyle." Warrendale, Pa.: Society of Automotive Engineers. (SAE Paper No. 801370.)

Energy and Environmental Analysis, Inc. 1980. *Medium- and Heavy-Duty Truck Fuel Demand Module Update and Calibration of the Highway Fuel Consumption Model*. Arlington, Va.: Energy and Environmental Analysis, Inc. (U.S. Department of Energy contract no. DE-AC01-79PE-70032, task 5)

Executive Office of the President. 1981. "Actions To Help the U.S. Auto Industry." The White House, Office of the Press Secretary. April 6.

Jambekar, A. B., and J. H. Johnson, 1981. "Effect of Truck Dieselization on Fuel Usage." Warrendale, Pa.: Society of Automotive Engineers. (SAE Paper No. 810022.)

National Research Council. Forthcoming. *Diesel Technology*. Diesel Impacts Study Committee, Technology Panel. Washington, D.C.: National Academy Press.

Society of Automotive Engineers. 1981. *Alternate Fuels*. Warrendale, Pa. (SP-480)

Tierney, W. T., et al., 1975. "The Texaco Controlled Combustion System, a Stratified Charge Concept: Review and Current Status." Paper presented at the Power Plants and Future Fuels Conference, Institution of Mechanical Engineers, January.

U.S. Environmental Protection Agency. 1980a. "Control of Air Pollution from New Motor Vehicles and Motor Vehicle Engines: Gaseous Emission Regulations for 1984 and Later Model Year Heavy-Duty Engines." (Final rule.) *Federal Register* 45(14):4136. January 21.

_____. 1980b. "Draft Regulatory Analysis, Environmental Impact Statement and NO_x Pollutant Specific Study for Proposed Gaseous Emission Regulations for 1985 and Later Model Year Heavy-Duty Engines." Office of Mobile Source Air Pollution Control. Washington, D.C.: U.S. Environmental Protection Agency. November 25.

_____. 1981a. "Control of Air Pollution from New Motor Vehicle Engines: Certification and Test Procedures." (Notice of intent.) *Federal Register* 46(70):21628. April 13.

_____. 1981b. "Control of Air Pollution from New Motor Vehicles and New Motor Vehicle Engines: Particulate Regulation for Heavy-Duty Diesel Engines." (Proposed rule.) *Federal Register* 46(4):1910. January 7.

_____. 1981c. "Control of Air Pollution From New Motor Vehicle Engines: Gaseous Emission Regulations for 1985 and Later Model Year Light-Duty Trucks and 1986 and Later Model Year Heavy-Duty Engines." (Advance notice of proposed rulemaking.) Federal Register 46(12):5838. January 19.

Appendix A
SUMMARY OF COMMITTEE CONTACTS

SITE VISITS

o   Environmental Protection Agency, Ann Arbor, Michigan, December 17, 1980.

o   Ford Motor Company, Detroit, Michigan, February 4, 1981.

o   General Motors Corporation, Detroit, Michigan, February 4, 1981.

MANUFACTURERS' BRIEFINGS

A number of manufacturers were contacted by letter and asked to provide written answers to the questionnaire shown in Appendix B. In addition, they were requested to present oral summaries of their submissions to the committee in Chicago on January 8, 1981. These manufacturers were selected by the committee as the ones the committee could learn most from, especially in diesel technology, as the starting point for our study. Each manufacturer gave a separate, confidential presentation.

The manufacturers were:

o   Caterpillar Tractor

o   Cummins Engine Company

o   General Motors Corporation

o   International Harvester Company

o   Mack Truck Company

o   Mercedes-Benz of North America, Inc.

VISITS AT SOCIETY OF AUTOMOTIVE ENGINEERING MEETING, DETROIT, MICHIGAN
FEBRUARY 23-27, 1981

o   Aldo Fozatti, Fiat of North America, Dearborn, Michigan.

o   Hans List, AVL Company, Graz, Austria.

o   Godfrey Greeves, Lucas-CAV, Ltd., Acton, London, England.

o   Franz Pischiner, University of Aachen, Aachen, West Germany.

o   Klaus Zimmerman, Robert Bosch Company, West Germany.

o   John J. Mooney, Technical Director, Engelhard Industries Div., Engelhard Minerals and Chemicals Corporation, Edison, New Jersey.

Appendix B
LETTER AND QUESTIONNAIRE

December 8, 1980

Dr. Laurence S. Caretto's
Invitation Letter sent to Automotive
Heavy-Duty Manufacturers with
NO_x Membership List & Questionnaire

Dear

The National Research Council of the National Academy of Sciences has formed a committee to investigate the technological feasibility of industry meeting the 1985 Nitrogen Oxides Standard for heavy duty vehicles (both gasoline and diesel). This committee, whose membership is shown in Attachment A, will meet in Chicago on Thursday, January 8, 1981. The specific purpose of this meeting is to discuss industry's current and planned activities for developing heavy-duty engine systems with low nitrogen oxides emissions. We invite you to make a presentation on behalf of Mercedes-Benz of North America, Inc.

The committee is working on a compressed schedule, which requires the completion of its final report by the middle of April. Our deliberations will directly impact the Environmental Protection Agency's adoption in June 1981 of a heavy-duty vehicle nitrogen oxides emission standard for 1985. We believe information gathered at the January 8th meeting will constitute a significant part of our study. This meeting represents the committee's initial contact with the specific concerns of the heavy duty engine manufacturers and suppliers regarding control of NO_x emissions. We are planning future contacts with the various concerned organizations and we would like to make site visits to selected manufacturers. Thus, the January 8th meeting will provide you with an opportunity to discuss your company's engine/control technology systems development activities with the entire committee.

The specific questions the committee would like you to answer for the January 8th meeting are shown in Attachment B. Unfortunately, we will have a very limited time to hear your oral presentation. Therefore, we would appreciate receiving, prior to the meeting, written responses to our questions and supporting documentation wherever possible. If you request, all materials will be held in strictest confidentiality. We request you send to the meeting members of your engineering staff, who will be able to answer the committee's technical questions.

The meeting schedule will be arranged so that only one company at a time will be present and give information to the committee. We hope you will be as candid as possible with the committee, so we may fully evaluate your company's potential for complying with heavy duty NO_x emissions standard over both the short (1-5 years) and the long terms.

Dr. Laurence S. Caretto's December 8, 1980
Invitation Letter Page Two

    The meeting will be in the O'Hare Hilton Hotel at Chicago's O'Hare Airport. Questions on scheduling arrangements should be addressed to Dr. Dennis F. Miller or his secretary, Vivian Scott (202/389-6974). A block of rooms has been set aside for those attendees requiring overnight accommodations. Therefore, if you need such accommodations, please telephone Ms. Scott. We will contact you to confirm your attendance and to give you a specific time for your presentation. We will also need to know whether you require audio or visual equipment for your presentation.

    I am looking forward to seeing you at this meeting.

                                           Sincerely,

                                           Laurence S. Caretto
                                           Chairman
                                           Motor Vehicle Nitrogen
                                           Oxide Standard Committee

LSC/vs
Attachments

cc: Dr. D. F. Miller/NAS

Attachment B — Questionnaire

HEAVY-DUTY NITROGEN OXIDES STUDY

Please provide answers to the following questions with as much detail as possible.

1. Describe your current line of heavy-duty engines (gasoline and diesel). Give the number of units sold and the emissions characteristics of each line.

2. Describe the engine/control-system packages that you have under development for possible sales in both the short (1-5 years) or long (beyond 5 years) terms. For each system, give specific engine and/or vehicle test data, and the tradeoffs for each on the following items. (Please note the test cycle used):

    (a) Gaseous emissions of regulated pollutants, e.g., hydrocarbons, carbon monoxide and nitrogen oxides;

    (b) Particulate emissions;

    (c) Emissions of unregulated pollutants, e.g., aldehydes, and odor;

    (d) Fuel economy and fuel economy penalty, if any, due to emission control;

    (e) Engine power output and any reduced output caused by emission control design features;

    (f) Other engine performance characteristics you consider significant, e.g., driveability, reliability, maintenance requirements, etc.;

    (g) In-use performance and deterioration of emission controls;

    (h) Costs attributed to emissions control features; this includes initial capital costs, changes in operating and maintenance costs, and fuel costs;

    (i) Manufacturing difficulties, if any, and lead-time requirements for production of the control system; and

    (j) Current development status and most likely and earliest possible introduction dates.

3. Describe work in progress, if any, on alternatives of engines such as gas turbines, adiabatic diesel, etc. What are emission and control problems with such engines?

Attachment B

4. Do you have any data concerning the effect of fuel properties on emissions? Are you testing alternative fuels, such as methanol, in your engines?

5. What engine designs do you plan to market over the next ten years? What are your estimated production figures for each type?

6. Do you have any information on the relative contribution of heavy-duty engines to total mobile source emissions?

7. List data on performance of specific after treatment devices you are considering which can be analyzed separate from the engine system.

8. What comments do you have on the test procedures for heavy-duty engines (gasoline and diesel) and the standards that can be achieved for heavy-duty engines?

Thank you for your help in providing answers to these questions.

Appendix C
CERTIFICATION TEST CYCLES FOR HEAVY-DUTY ENGINES

This appendix provides a brief description of the test procedures used in determining the emissions from heavy-duty diesel and gasoline engines. The procedures in current use are scheduled to be replaced in the 1984 model year (U.S. Environmental Protection Agency, 1980). As the technical discussions in this report make clear, the change is of substantial significance.

The current "steady-state" procedures involve measurements of emissions while the engines are operated on a dynamometer at precisely defined sequences of steady operating states. The test cycle for diesel engines consists of 13 such states, each a combination of specified speeds and loads; the cycle for heavy-duty gasoline engines consists of a similar series of 9 operating states. These test cycles are commonly called steady-state cycles.

For 1984 and beyond, heavy-duty engines will be tested for emissions on new "transient" test cycles, which involve complicated but precisely specified series of accelerations and decelerations at specified changing conditions of load and speed. Again, diesel and gasoline engines will be subjected to separate tests.

To accommodate the variety of heavy-duty engines with a single test cycle for each of the two engine types, the cycles are defined in terms of percentages of the maximum observed torque at each speed. Two different engines, each with its own performance characteristics, would thus be tested at different torque values at the same points in the cycle, but both would have the same value for the percentage at each point in the cycle.

Figure C-1 shows the steady-state test procedure for gasoline engines. This is a nine-mode cycle, performed two times. The engine speed is kept at 2000 rpm, except for idle conditions when the normal engine speed is used. The four percentage torque points (10, 25, 55, and 90 percent) correspond to part-throttle deceleration, cruise, part-throttle acceleration, and full load. The other operating modes are specified as idle and closed-throttle deceleration.

The steady-state cycle for diesel engines is shown in Figure C-2. In this cycle the engine is run in the idle mode and at specified torque

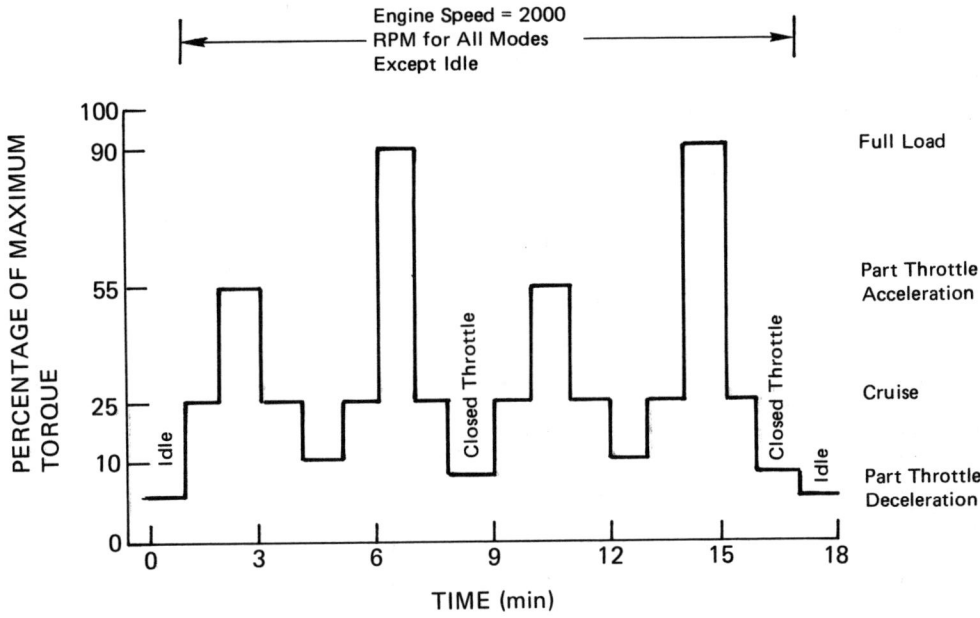

Figure C-1 Steady-state test cycle for gasoline engines

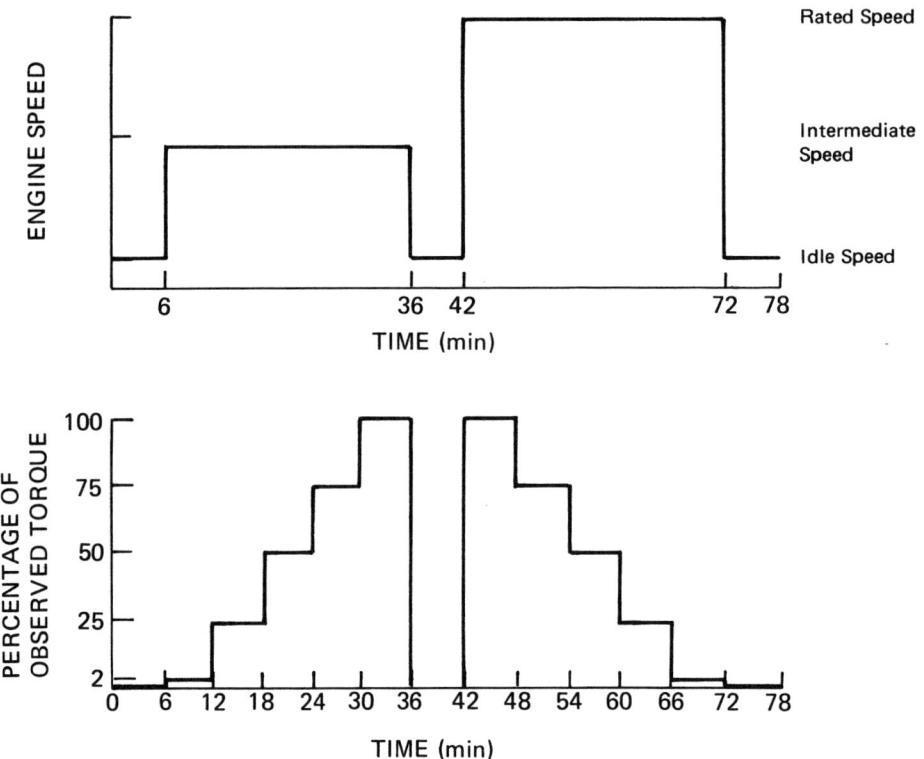

Figure C-2 Steady-state test cycle for diesel engines

percentages for two engine speeds, the maximum rated speed and an intermediate speed. Although the percentage torque values are the same at both engine speeds, the actual torque is different because the maximum torque is different at the two speeds.

In the steady-state cycles emissions are measured only during the specific operating modes. Emissions resulting from changes between the different modes are not considered.

In the transient test procedures, both the speed and the torque are specified as percentage values. Prior to running the transient cycle it is necessary to measure the maximum and minimum engine speed and to obtain a graph of maximum torque as a function of engine speed. These data are then used to convert the cycle specifications into actual speed and torque values for the specific engine.

The transient test cycle for a Volvo TD100C diesel engine is plotted in Figure C-3. For other engines the actual speed and torque values would be different, but the shapes of the curves would be the same. The notations at the top of Figure C-3 indicate the types of vehicle operation used to generate the cycle; these are New York nonfreeway (NYNF), Los Angeles nonfreeway (LANF), and Los Angeles freeway (LAF). The negative torque valves shown in the figure simulate conditions in which the engine is being driven by vehicle inertia (e.g., deceleration). Manufacturers that do not have dynamometers capable of driving engines to the negative torque levels must buy new dynamometers to perform the new test procedures.

The power output of the engine during the various portions of the cycle can be visualized by recalling that power output is the product of torque and speed. Thus regions where both torque and speed are high are regions of high power output, and, regions where both are low are regions of low output.

Figure C-3 is intended to illustrate the types of engine operation required by the transient test cycles. The cycle for gasoline engines has a different set of specifications, but the general nature of the cycle is similar. The full details of both test cycles are given in U.S. Environmental Protection Agency (1980).

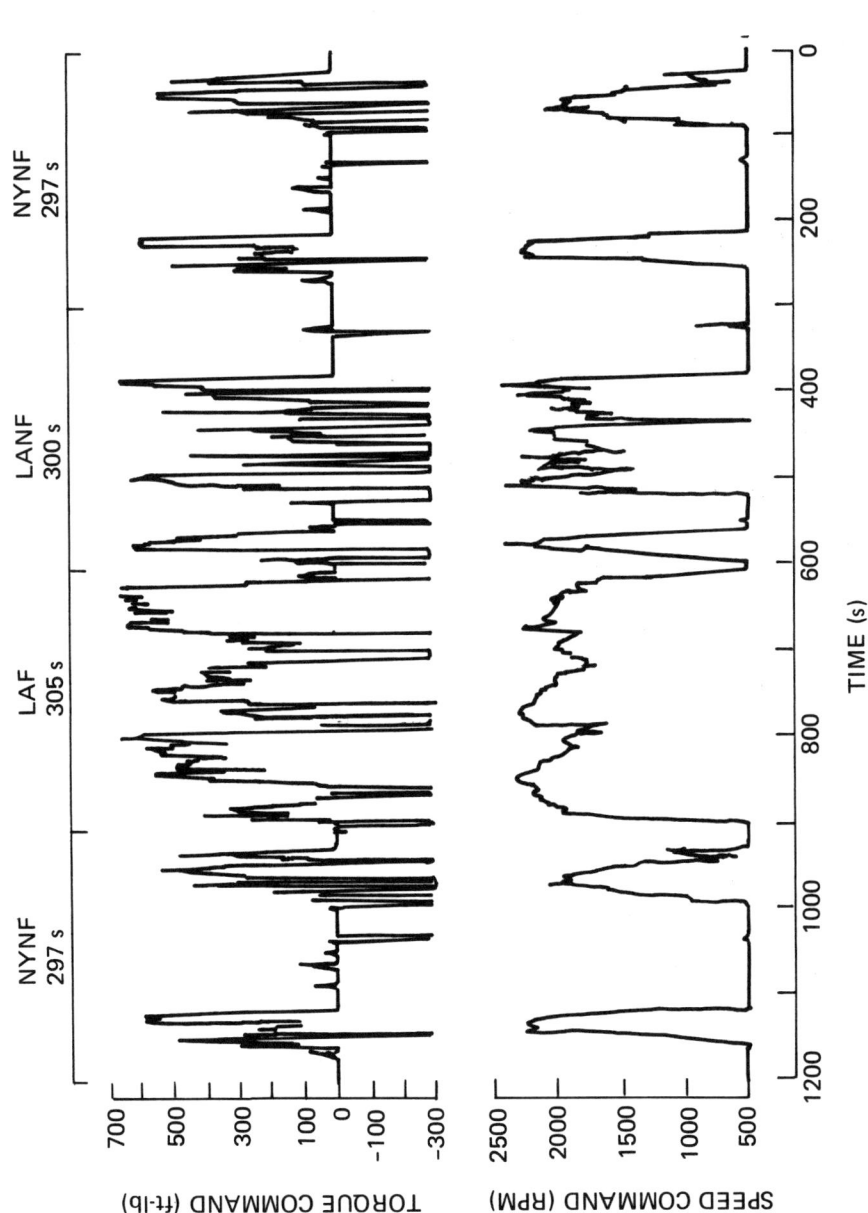

Figure C-3 Graphic representation of torque and speed commands in the transient test cycle for diesel engines, based on a power map of the Volvo TD100C (Ullman and Hare, 1981)

REFERENCES

U.S. Environmental Protection Agency. 1980. "Control of Air Pollution From New Motor Vehicles and Motor Vehicle Engines; Gaseous Emission Regulations for 1984 and Later Model Year Heavy-Duty Engines." (Final rule.) Federal Register 45(14):4136. January 21.

Ullman, Terry L., and Charles T. Hare. 1981. "Emission Characterization of an Alcohol Fueled-Diesel Pilot Compression-Ignition Engine and its Heavy-Duty Diesel Counterpart." Draft report prepared for U.S. Environmental Protection Agency, Office of Mobile Source Air Pollution Control, Emission Control Technology Division, Ann Arbor, Mich. (Contract No. 68-03-2884, Task 6.) June.

APPENDIX D
GLOSSARY OF TECHNICAL TERMS

AIR/FUEL RATIO: Ratio of air to fuel, by weight, in a combustion process.

BRAKE-SPECIFIC QUANTITY: The prefix "brake-specific" denotes a measure of some quantity divided by the power delivered by the engine. The total power generated in the engine is called the indicated power. The net or brake power output of the engine equals the indicated power minus the power required to overcome engine friction.

CATALYTIC CONVERTER: Device in the exhaust systems of vehicles with spark-ignition engines, containing a catalyst to aid in reducing emissions. (See oxidizing converter and three-way converter.)

CETANE INDEX: Value dervied from an empirical method for finding the cetane number of a diesel fuel, based on fuel's density and mid-boiling point.

CETANE NUMBER: Equivalent percentage by volume of cetane (= n-hexadecane, cetane number 100) in a mixture with alphamethylnapthyene (cetane number 0); indicates the relative ignitibility of the fuel upon injection into the engine cylinder.

COMPRESSION RATIO: Ratio between the volume above the piston when it is at bottom dead center to the volume above the piston when it is at top dead center.

DYNAMOMETER: Device for testing engines. Assures power output from the engine and provides measurement of engine torque.

EGR: Exhaust gas recirculation. Recirculation of a portion of the engine exhaust gases into the air intake, to control NO_x emissions.

ENGINE LOAD: General term referring to power or torque demand on the engine.

ENGINE SPEED: Crankshaft revolutions per minute.

FOUR-STROKE CYCLE: Engine cycle consisting of a downward intake stroke, an upward compression stroke, a downward power stroke, and an upward exhaust stroke.

HEAVY-DUTY: Applications more severe than passenger car service. Usually applied to service in buses or trucks in excess of 8,500 pounds gross weight rating.

HYDROCARBONS: Substances whose molecules are made solely from hydrogen and carbon atoms, with multiple carbon atoms linked to one another and with hydrogen atoms attached to the carbon atoms.

MANIFOLD: Branched-pipe passage device that connects openings from each cylinder to a common opening.

OXIDATION CATALYST: Catalyst, operating with oxygen in the exhaust gas, designed to oxidize exhaust carbon monoxide and hydrocarbons to carbon dioxide and water.

PERFORMANCE: General descriptive term covering engine power output, durability, and driveability.

QUENCH: Cool to temperatures at which the flame will go out.

SAC VOLUME: Volume of the space in the top of a diesel fuel injector. Zero volume is the ideal design objective for low hydrocarbon emissions.

SPARK-IGNITION ENGINE: An engine whose combustion is started by an arc across the spark plug electrodes.

SPARK TIMING: Point in an engine's crankshaft rotation at which the spark plug is fired.

STEADY STATE: Constant operating conditions with no variation in speed or load.

STEADY-STATE TEST: Current test cycle for heavy-duty engines.

TEST CYCLE: The specific sequence of speed and load conditions at which an engine is run for measurement of emissions.

TEST PROCEDURES: Full set of procedures for testing engines to determine compliance with emission standards.

THERMAL EFFICIENCY: Energy output (of the engine, for example) divided by the energy in the fuel used.

THREE-WAY CATALYST: Exhaust treatment catalyst designed for simultaneous removal of carbon monoxide, hydrocarbons and oxides of nitrogen.

TORQUE: Measure of the twisting moment on a shaft.

TRANSIENT TEST CYCLE: The new test cycle for heavy-duty engines, effective in 1984.

Appendix E
STATEMENT OF GEORGE R. HEATON, JR.

While I endorse the technical findings of the Committee, I disagree with some elements of the approach that it adopted toward the questions at hand. Moreover, since I believe that reports such as the Committee's inevitably constitute "policy" statements, an additional comment on the document's policy relevance seems to me to be in order.

The Committee chose to confine its inquiry largely to "technical" matters. Thus, the report deliberately avoids explicit comment on the policy choices faced by EPA and the Congress. In retrospect, it appears to me that the Committee focused principally on the state of development of NO_x emissions control technology. Accordingly, the two chapters on gasoline and diesel engines comprise the bulk of its report. In my view, the paramount question was: what heavy-duty vehicle NO_x requirements should be promulgated under the Clean Air Act? Had this been the Committee's focus, a different kind of report would have emerged, emphasizing more heavily the complex health, economic, and policy design questions involved, without excluding the technical component.

Naturally, the Committee was limited by time and resources. Given these constraints, it concentrated its efforts where it believed they would be most productive. I simply disagree with this choice, for two main reasons. First, I believe that under the Clean Air Act, this committee was given the authority and responsibility to comment on the policy determinations inherent in the heavy-duty vehicle NO_x rulemaking. Its decision not to do so deprives EPA, the Congress, and the country of valuable information, analysis, and expertise in dealing with a complex regulatory issue.

Second, although the Committee thought it could avoid the policy issues, I believe that in fact it could not, and thus, would have done better to address them directly. As it turns out, the report, both in its general orientation and factual findings about emissions control technology, strongly supports the position EPA has taken to date. Implicitly, the Committee's work contains the following policy conclusion: the 75-percent reduction in NO_x emissions mandated under the Clean Air Act is not feasible whereas the level now apparently intended by EPA (4 g/bhp-h) is.

I also feel that within the confines of its technical assessment of NO_x emissions control technology, the Committee focused too much on near-term development taking place among the engine manufacturers. Recognizing that the Committee saw the 1986 time frame as its context, I nevertheless believe that some broadening of focus would have been appropriate. Technology development is a dynamic process. What occurs in the near term is strongly dependent on how firms see the future, particularly the future of regulatory requirements. If little is required, most certainly little will be developed. The uncertainties involved in technology development to comply with regulation perhaps fall most heavily on supplier firms; nevertheless, these firms have been an important source of innovation in the past, as have foreign firms and new entrants to an industry. In my view, the Committee might have given more consideration to these broader long-term possibilities. The failure to do so tends to lead one to the policy conclusion that strict controls on NO_x emissions are not technologically feasible.

Going beyond what the Committee has done, my own view is that EPA has too easily and without sufficient analysis abandoned the 75-percent emissions reduction goal in the Clean Air Act. Of course, regulatory decision-making of this sort is fraught with uncertainty, and one cannot be sure about what control technology will be used, or what its cost will be. While others may reasonably resolve their doubts differently, I would resolve mine in favor of the statutory goal. Seeking the strict standard need not have harsh economic effects. Different standards could be set for diesel and gasoline engines, for example, or the time period for compliance might be relaxed. Moreover, a failure by the regulators to adhere to congressional goals can adversely affect the firms that have led in the development of technology.

The principal doubt that overlies the whole question of NO_x control concerns the diesel engine (i.e. the available control technology, the health effects trade-off between particulates and NO_x, and the fuel economy implications). Its resolution could cause my views to change. Not enough is known now about this issue, and more analysis should be done.

Lastly, although there are aspects of the Committee's approach and report with which I disagree, I cannot fault the process by which the Committee reached its conclusion. The issues raised in my statement have been aired in the Committee and during the report review period, and the majority of the Committee simply had a different view from my own.